Toy Design

Irene Guerrieri

Toy Design

From Concept to Realization. A Guide for the Designer

Springer

Irene Guerrieri
Sarnico, Bergamo, Italy

ISBN 978-3-031-98871-4 ISBN 978-3-031-98872-1 (eBook)
https://doi.org/10.1007/978-3-031-98872-1

The original submitted manuscript has been translated into English. The translation was done using artificial intelligence. A subsequent revision was performed by the author(s) to further refine the work and to ensure that the translation is appropriate concerning content and scientific correctness. It may, however, read stylistically different from a conventional translation.

Cover illustration: Fork-ship, Traffic line, Fratelli Guzzini (2017), design Irene Guerrieri

This Springer imprint is published by the registered company Springer Nature Switzerland AG
The registered company address is: Gewerbestrasse 11, 6330 Cham, Switzerland

To my grandmother Luisa,
for being my
first guide to drawing

Preface

This is a book born out of passion and love for the topics covered. Starting to write it was an opportunity for me to reorganize the topics I bring to my seminars, creating a logical sequence accompanied by some practical proposals that can help the reader follow a project path.

A small aid, a *vade mecum*, for those who want to immerse themselves in the fantastic world of toys.

I address adults even though my language and my project research are dedicated to children because the adults who venture into reading this and conducting the exercises proposed within will establish and improve dialogue with children as a crucial preliminary stage prior to any design.

Approaching toy design means, above all, learning to listen and to look at the beauty around us with the innocence, freshness, and curiosity of children.

A path that I love to define as liberating and that can give great satisfaction if you know how to put it into practice.

The beginning of any project activity lies in the observation of what surrounds us, which constitutes the well from which we draw our ideas. The world is full of beauties that not all of us are always able to look at and feel. Rodari teaches us that with a "child's ear" we can learn to "listen to what the trees, birds, passing clouds, stones, and streams" say. It takes training, curiosity, and attention.

You need to equip yourself with a notebook with white sheets of consistent weight, bound with a hard cover, easily transportable and openable that can also be well preserved over time, because it is with time that ideas develop. An observation made today, in fact, can give rise to a project tomorrow, and that's why it's important to note and preserve, to then resume, even at different times, the sketches made yesterday. Noting and sketching what intrigues us is the foundation for emerging ideas; it's best to do it with a pencil and some color, a method that, although it may seem old-fashioned and artisanal, I consider emotionally the most effective and cannot be replaced by any other means. Every project always starts from the encounter of a pencil with a piece of paper. The choice of manual drawing, as a primordial communicative language to express concepts, always remains the most powerful, as powerful as taking inspiration from the elements present in nature to find solutions to our ideas.

The aim of this small manual is to help the reader rediscover their own secret garden, sprouting the seeds that are inside us and that come from the outside world. The rule is to bring the outside in and the inside out, reconnecting with the child we once were and bringing him/her back to life. The things that surround us are the spark to ignite the imagination; each of us is a bit of a child and a bit of a poet.

Activating passions is essential, as without them, there can be no knowledge or existence, and nothing good can be created. The road that leads to a good project is much like the road to happiness, to the rediscovery of the child within, in the dialogue that takes place between the adult and his inner predecessor. A journey into memory, from which our creativity never ceases to draw.

To that end, it is my desire to provide some tools of the trade, personally tested, to help bring out the best in each person, reviving their inner child.

The toy project has a profound liberating charge as it connects our past with our present, making us aware that our creativity springs from the information we have captured from lived experiences, and the more we have, the more potentially creative we are.

In 1990 I met Bruno Munari for the first time in his studio in Milan, and I returned to visit him many times. Every time he told me interesting things; he spoke to me in a serious but fun way, so I quickly learned that communication is better when it is entertaining.

I also learned this lesson by reading the books of Gianni Rodari and other authors who, like him, understood that cultivating the humorous side of everything makes learning easier.

Munari argued that a child who lives in the countryside is potentially more creative than a child who lives in the city, because nature provides many more tips and gives good ideas, and experiences lived in contact with nature are experiences that remain indelible. From the forms of nature, considered as an inspiring source, many inventions arise; just think what numerous examples in architecture and design are inspired by the shapes and colors of nature.

But any element with which we come into relation, and that we like, can work as a magic object to unearth fields of memory if we know how to "listen" to it. Well, a toy must work as a magic object, where the role of the designer is to play with the elements that define it, finding the appropriate language.

Knowing how to play with the objects of memory to give life to the toy means having a great communicative power. All this requires training and practice. It is therefore advisable to repeat the exercises that accompany the text several times and in different contexts, and to obtain a good sketchpad to have ready as you start to read this book.

Because even imagination is a quality to develop and exercise.

Sarnico, Italy Irene Guerrieri

Competing Interests The author has no competing interests to declare that are relevant to the content of this manuscript.

Contents

1

Game and Toy

Two Concepts for the Same Activity

Abstract The toy is the object, and the game is the action that takes place using the object. What binds the two terms is the planned play activity. There are games that do not require objects and objects that instead establish playful activity. The ball is an object that can be played in many ways, so it needs to be designed differently depending on the game you want to play.

We may all agree in defining a toy as the object and a game as the action that takes place using the object, or perhaps in its absence. What binds the two terms is playful activity.

However, there are a series of issues to delve into that involve the designer in the process of bringing life to a new product intended for this type of activity.

Take, for example, the ball: a toy with spherical shape, to which the designer can attribute dimensions, materials, weight, construction details, colors, and graphics. The ball can give rise to an infinite number of games, and it is precisely based on the activity that we want to carry out with it that the designer will need to think about and decide on all its characteristics so that it can satisfy the intended functions.

It is not excluded that often the designers of a toy are also the creators and proposers of the games that can be realized with their product; therefore, the concepts of game and toy, which could seem independent, must consider each other.

I. Guerrieri, *Toy Design*, https://doi.org/10.1007/978-3-031-98872-1_1

There are many games available that don't require specific objects to play.

Regardless of their handcrafted or industrial production origin, toys function by bringing playful activity to life.

In common usage, the broader term "game" often encompasses the concept of a "toy." When materials have a high weight on the structure of the game itself, such as card games or board games, a stronger identification is created between the object and the type of game. We say, "Shall we play this game?" when in reality it's a "toy," that is, an object, but we don't call a deck of cards a "toy" unless they are used to build something, like a tower or a little house. Generally, the word "game" absorbs the term "toy" in all board games designed for an older age group.

In summary, we say that the "game" presupposes rules and can also be played without objects, while the toy corresponds to a physical object and can be used in multiple ways. So we can discuss "rules of the game" instead of "rules of the toy," since the toy's purpose is to follow the game's rules. Some toys, such as dolls, puppets, pinwheels, and trains, are meant for contemplation; in these cases, the action is minimal, but simply paying attention to the object can inspire imagination and fantasies.

Or there are toys that require more manual skills, with the exercise of simple and repetitive actions, where the fun consists precisely in learning to use them by making mistakes and developing skills, like the yo-yo, the hoop, the kite, the ball, etc.

In essence, the toy can be played with more freely, without rules. A toy may contain instructions for its operation or its assembly, but the playful and fantastic activity that follows must be something else and invest in the emotional, creative, and imaginative sphere of the child. The instructions that allow us to operate a toy should not exhaust the playful activity.

A toy, to perform its correct function, must provide possibilities for adventure, discovery, surprise, and joy. The designer must take this purpose into account.

To survive, a toy must entertain the child with its imagination. Through the relationship with the toy, children project their world. This relationship can also lead to the extreme act of breaking the toy, which can occur to discover its mechanisms, to express one's own aggressiveness, or simply because the object has exhausted its function.

There are two types of toys: those that are designed for children to play with, and those that are not originally toys but become so when given a role and meaning. In this second category, all those objects present in a house, not structured as toys, are included and become toys in the hands of the child. These are mostly common use and domestic objects because children have a particular propensity to transform everything they find into a toy, as play is the fundamental way in which they learn about the world. The broom becomes a horse to ride or a spaceship, the chair a little house under which to take refuge, the blanket a hut, and so on. These unstructured toys, however, have great importance from the cognitive perspective. The child has playful experiences of great value and develops imagination and creativity.

The primary task of the designer is to devise toys capable of providing meaningful and memorable playful experiences. Therefore, a good rule is to design toys without structure, allowing the child's imagination and creative contribution to roam freely. After all, as a rule, the designers themselves must listen and bring out their inner child, seeking a creative association of elements that come from different conceptual spheres but that, put together, give life to a new concept, fantastic and creative, in which the child recognizes the original elements and, retracing the path that led the designers to conceive that toy, enriches his own imagination with new solutions.

Designers capable of simplifying and making their project as unstructured as possible give the child the opportunity to continue to develop and enrich the project itself.

Modern pedagogy attributes great importance to unstructured toys for the education and cognitive development of the child.

The child–toy relationship must always be an educational relationship, and in this regard, we call the activity that emphasizes this function an "educational game."

Good designers must always keep in mind the educational purpose of their project, without which their products will mostly be decorative.

Having said this, play materials must help to develop certain skills and knowledge at an intellectual, sensory, and psychomotor level. Fundamental contributions have been provided by Friedrich Fröbel, Maria Montessori, and, as far as unstructured materials are concerned, by the Agazzi sisters.

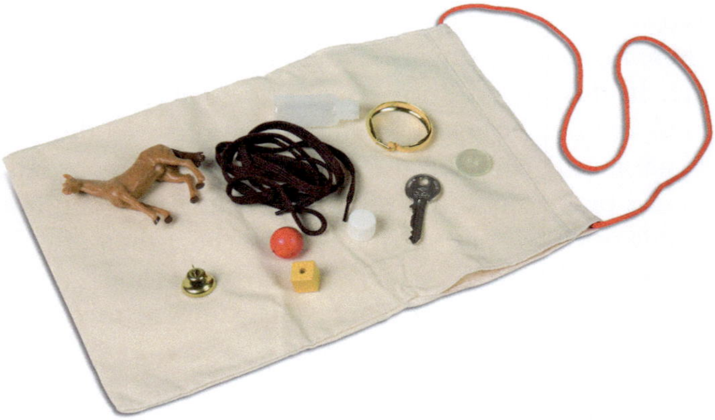

Unstructured materials, GAM Gonzagarredi Montessori

2

A Dive into Memory

There are Deep Echoes Within Every Object with Which We Interact

Abstract As designers and inventors of new products, we always must deal with our experience and with the data acquired by our 5 senses and collected from our memory, which is the fuel of our creativity.

Let's ask ourselves the question, "How is an idea formed?"

I will attempt to clarify some fundamental aspects of this phenomenon by discussing the cognitive activities that we commonly refer to as imagination, invention, and creativity.

Let's focus on the way our intelligence operates, taking into account the stored data received from the outside world, but also from our very personal inner world.

As designers and inventors of new products, we must always deal with our lived experiences. The richer, the more open, and the more receptive our experiences are, the greater the possibilities of developing and transmitting creativity. Not everyone possesses unique creativity or a wealth of experiences that foster the development of creative mechanisms.

The outside world is explored, captured, and stored by intelligence, which, through logical operations, allows us to understand the things around us.

This operation takes place through the senses, as we receive everything that surrounds us through sight, hearing, touch, taste, and smell. What is received becomes fixed in memory.

I. Guerrieri, *Toy Design*, https://doi.org/10.1007/978-3-031-98872-1_2

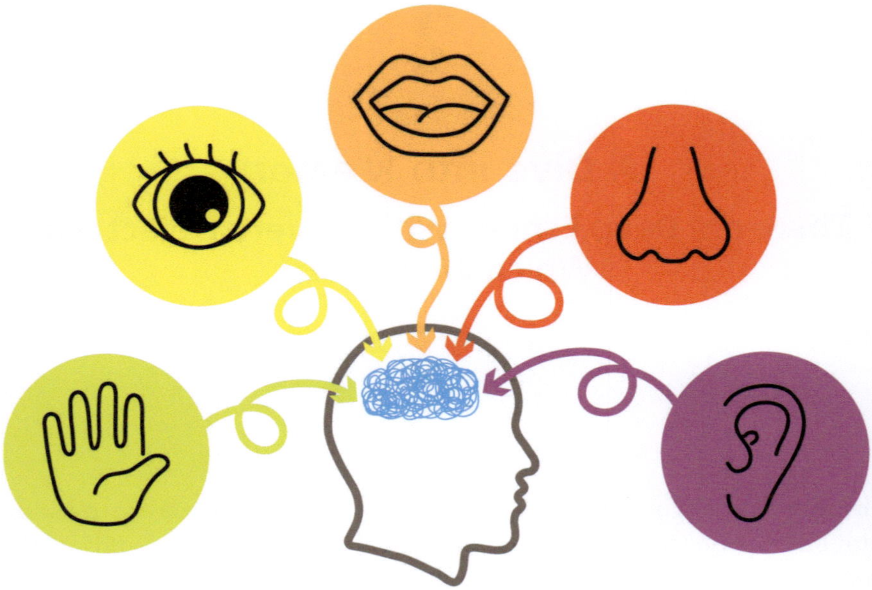

Our brain receives information from our senses

It is evident that a child's memory contains a limited amount of information, whereas that of an adult is filled with a plethora of information. Additionally, the memory of an individual who has encountered a multitude of experiences and observations will undoubtedly be more extensive.

We receive deep echoes from everything with which we interact because it stimulates a memory, causes us to relive events and feelings, or simply recalls an idea.

Memory plays a central role for the designers in their creative process.

We gather a wealth of information from the outside world through our senses, which intelligence encodes and stores in our memory. From this part of our brain, where memory resides, imagination, invention, and creativity spring forth, which can or cannot be visualized only through imagination, that is the visualization and design of an idea. Once the design process makes them visible, we can realize them as products, further enriching the outside world.

Often, based on experiences and echoes received from objects diametrically far from what we are about to conceive, we rework information that can bring creative, inventive, or fantastic utility to our project. From memory spring new ideas, reworking and mixing received concepts.

Even if it is unachievable, something that we produce with our head can be a fantastic idea. Imagination has no constraints and is free to think even the most absurd and impossible things. Children are especially capable of this when they conceive fictional characters as original combinations of objects, animals, people, and natural phenomena.

An invention is always rooted in our memory, as it considers acquired data that enables the conception of new ideas; however, it focuses solely on practical applications and disregards aesthetic concerns. The invention aims for practical use. New mechanisms, new materials, and new software are products of the invention, which is unconcerned with the aesthetics of the object; only utility is of interest.

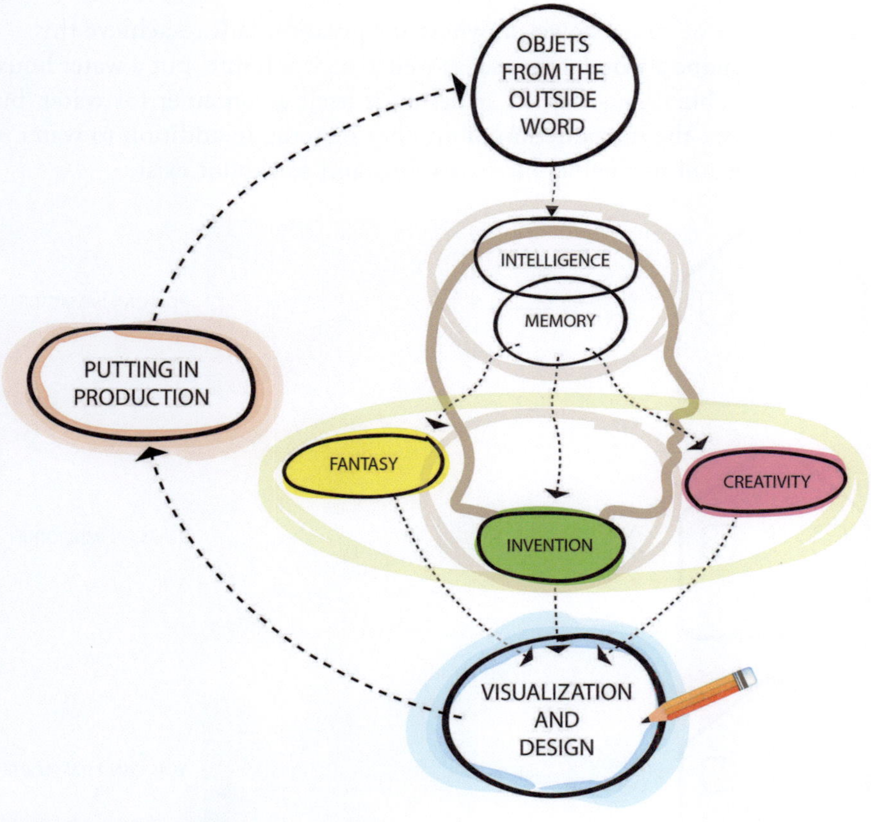

How our brain works

A creative idea, in contrast, is something that is aesthetically imaginative and intended for practical use. Design, including toy design, allows for the concrete realization of a product of creativity, with a targeted use of fantasy and invention. Design not only contains the aesthetics of fantasy and the utility of invention, but also takes into account the psychological and social aspects, and, in the specific case of toy design, also takes into account the educational aspect.

Visualization renders perceptible what we have imagined through fantasy, invention or creativity. We realize our imagination through drawings or, more effectively, through projects that bring the idea to life. Some people have less capacity of imagination and need assistance from professionals to visualize an idea. A model can also make an idea visible. An idea can be manifested in many different ways and styles; this depends on the designer's sensitivity, experience, and tools available, as well as the specific requirements for the project.

There are some cases, however, where imagination fails to achieve this.

We can imagine a wooden house, as well as a glass house, but a water house is not conceivable unless, say, its structure is itself a container for water, but this presupposes the introduction of another material in addition to water. A water house is unimaginable, inconceivable, and so cannot exist.

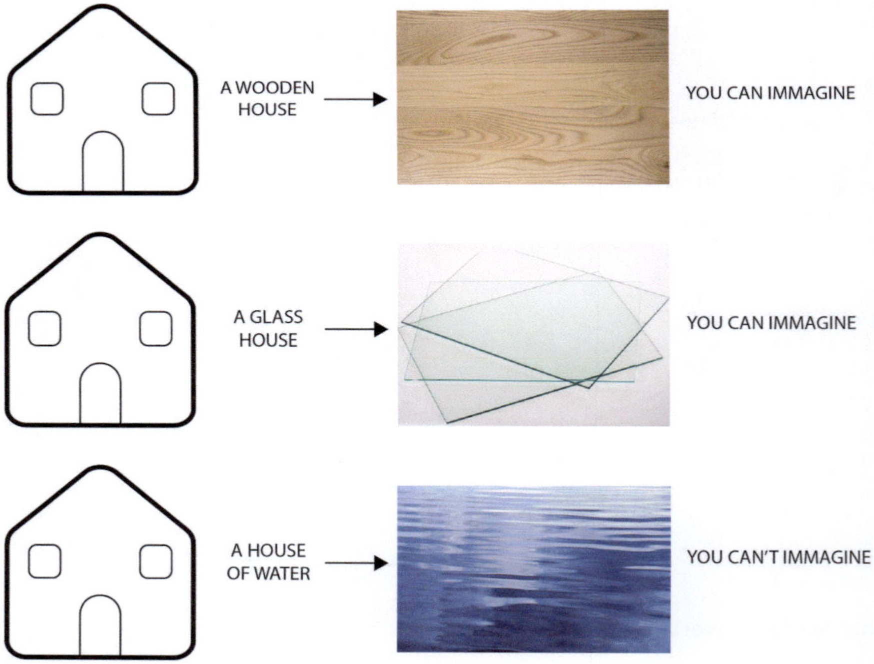

Imaginable or unimaginable relationships

Therefore, for an idea to become feasible, it is essential that it can be imagined and made visible.

A water house is a beautiful example of imagination, which is exciting, but it is such an absurd thing that it remains a fantastic idea to tell in words, but we cannot visualize it, because to imagine it, we must introduce a container capable of shaping the water.

For productive purposes, it is essential that every good idea can be visualized.

Once accepted, a visualized concept is put into production and subsequently returned to the outside world. This enriches the environment by introducing a new product and idea, thereby expanding the source from which our intellect is derived. Exponentially, the human community is enriched with new ideas, receiving new stimuli to produce more.

Everything revolves around memory in an existential cycle devoid of exhaustion.

3

Relationships Between Objects of Memory

The Fantastic Theme Arises When Unusual Combinations are Created

Abstract The inventions born from the meeting of concepts belonging to different categories. The fantastic theme arises only by the relationship between two distant concepts. The inventor and the designer find the relationships between acquired data and try fantastic relationships between known objects to give life to the new ones.

The fantastic theme arises when unexpected combinations are created, that is, when a relationship is established between objects that belong to distinct categories.

I. Guerrieri, *Toy Design*, https://doi.org/10.1007/978-3-031-98872-1_3

FISH + WHEELS

ANIMAL + FUR

CUBE + NUMBERS

SPOON + TOY CAR

FORK + SHIP

***Blubby pull toy*, children's line, Thun (2006)—design by Irene Guerrieri**

***Sheep*, pull toy line, *Zookids*, Sevi/Trudi (2002)—design by Irene Guerrieri**

***My first numbers*, line *Eco-cubotti*, Sassi Junior (2015)—design by Irene Guerrieri**

Spoon-toy car, line *Traffic*, Fratelli Guzzini (2017)—design by Irene Guerrieri

Fork-ship, line *Traffic*, Fratelli Guzzini (2017)—design by Irene Guerrieri

Therefore, a single element is not enough to arouse the fantastic theme; at least two are needed.

The single object only acts when it encounters a second object that provokes it, forcing it to step out of its usual behavior and reinvent itself in a new and stimulating form of communication.

Gianni Rodari writes in his *Grammar of Fantasy*: "It is not enough to have one electric pole to spark; there must be two. The single word acts only when it encounters a second one that provokes it and forces it to step out of the tracks of habit to discover new capacities to signify." What Rodari writes about writing we can say for design.

There is no life if there is no struggle; nothing is born from quiet, but from the clash and comparison of different themes belonging to distinct categories. The fundamental element of thought must be the pair: two objects give rise to a new object never seen before, unique in its communicative strength. Paul Klee, in his *Theory of Form and Figuration*, writes, "The concept is impossible without its opposite. There are no stand-alone concepts, but as a rule there are binomials of concepts." To produce a good idea, a fantastic binomial is needed, created by two objects between which there is a certain distance, so that their juxtaposition is decidedly unusual and forces the imagination to establish a kinship between them.

Exercise 1

In the two columns below, list a series of unusual combinations. Repeat this exercise several times, in different moments and situations.

Object 1	Object 2

We choose combinations: some are to be discarded, others stimulate the imagination, and others still give rise to real inventions.

You cannot relate things that you do not know. Generally, we explore objects by finding relationships with other objects that our memory brings to the surface. For example, a relationship can be established between an armchair and a baseball glove; thus was born *Joe*, a project by De Pas, D'Urbino, and Lomazzi.

If instead we decide to establish a relationship between two materials like paper and rubber, a stretchy sheet or a paper rubber band might result. Both are beautiful inventions!

***Joe Armchair* (1970)—De Pas, D'Urbino, Lomazzi, *Joe*, Armchair, Poltronova, photo Pietro Savorelli, courtesy of Centro Studi Poltronova Archive**

Establishing relationships, in life as in projects, is a fundamental step in creating new things.

In 2002, the *Zookids* line was born, which I conceived and designed for the Sevi brand, recently acquired by the Trudi company. I was asked to develop a line that would establish a connection between the two historic brands, which are both deeply rooted in the collective imagination and are associated with the two emblematic materials that the two companies use: the characteristic wood of Sevi and the plush that identifies Trudi.

The challenge for me was to relate the identifying materials of the two brands in an attractive and innovative way. The result was a line of thirty-six items developed on the theme of animals. Animals with a wooden silhouette and removable plush fur. The playability of the proposal lay in being able to dress and undress the animals, but also in exchanging furs between different animals. Other toys belonging to the same line were characterized by the use of wood and fabric with distinct functions attributed to the two materials: wood for the animal's form and fabric for the playable detail. The line included pull-along toys with wheels, animal puzzles, booklets with wooden and fabric pages, merry-go-rounds, and inset games. The *Zookids* line was selected for the ADI Design Index 2002 and published in important industry magazines.

It was a participant in the 2008 exhibition "D come Design" in Turin for the World Design Capital, which was subsequently presented in Cairo. It also participated in the 2016 exhibition "W women in Italian Design" at the XXI Triennale di Milano, as well as other exhibitions.

The more relationships we engage in, the more opportunities we have to unleash our imagination and create new things. The more we know about the reality that surrounds us, the greater the possibilities for making relationships.

In the child, who knows little, there is a projection of everything he knows and experiences firsthand onto things he doesn't know very well, but to which he attributes his own qualities and characteristics. The big carrot becomes the mother of the small carrot, and on the carrot is projected all the child's lived experience. Therefore, in children, we cannot properly speak of imagination but of projection. Commonly, however, we are used to saying that children have a lot of imagination; in reality, they project their lived experience on the objects they encounter and create fantastic, unexpected, and also very funny analogies.

***Zebra* pull-along toy, *Zookids* line, Sevi/Trudi (2002)—design by Irene Guerrieri**

"Once upon a time there was a snake that never wanted to go to sleep, and his bed was always well made…." Of course, a snake that doesn't want to go to sleep has never been seen, nor has a snake that even has a bed, but in children's imagination, attracted by the animal, there is a projection of what closely concerns their own life onto a snake that always wants to stay awake, just like them.

If we want the children to develop their imagination, we must introduce them to many things and allow them to participate in a wide range of activities to help them gain experiences and accumulate information. A creative person with a developed imagination can solve many problems, from the simplest, related to everyday practical life, to the complex.

At the base of this process is the ability to interrelate things. The increase in knowledge allows a greater number of relationships between a larger volume of acquired data. In some cases, the relationships are only used to understand a concept or to obtain a result—for example, two plus two equals four—and in other cases, relating things brings about fun, play, and also (indeed, why not?) a healthy laugh.

Exercise 2

Repeat Exercise 1, listing combinations that bring about a healthy laugh and stimulate fun.

Object 1	Object 2

Let's now try to establish relationships between objects and their qualities. We are assigning a quality to an object that typically does not belong to it. For example: a "giant" "ant," a "flat" "house," "Red Riding Hood" "blue," "salty" "honey,"… and so on.

Exercise 3

Write a list of things, animals, or people with mismatched attribute next to it, which may concern its shape, its color, its weight, the material, or its meaning.

Object 1	Attribute

In this chapter, I tried to explain the importance of being able to relate things, even things belonging to different categories. It is not enough to memorize them; you must also relate them. Finding connections requires more than just memorizing information.

In this, children are giants. As previously mentioned, children naturally project their small world into everything they encounter, establishing relationships.

Adults usually have more data, but they may not always find good connections amongst it. So, an adult is only potentially more advantaged because he knows more things and, over the years, has memorized more data, but if he does not return to some notion of childlike thinking and resume the connection activity that he was able to do with projection, he is unable to produce any new ideas.

As adults, it is our responsibility to assist children with expanding their knowledge and memorizing increasing amounts of facts, and we must do so in a joyful and engaging manner through play.

This is an endless cycle from which new ideas occasionally emerge, capable of enriching collective knowledge. Included in this process is the duty of those who design toys and want to repay to the "well" from which they have drawn, new aids for learning and development.

There are many things in nature that remind us of others; for example, a heart-shaped leaf or a stone shaped like a face, and many more. Observing these things and noticing hidden relationships behind natural appearances trains our imagination and ability to find connections.

Exercise 4

Look around and choose an object that you see that attracts your attention and makes you imagine something else, then write it here.

Object 1	What it makes me think of

Another useful tool for creating relationships is the metaphor: commonly used in writing, but it can also be used in the design of objects and toys to strengthen their message.

What is a metaphor? A metaphor is a rhetorical figure that replaces one term with another based on an implied similarity. We use it to enhance comprehension and vividness of our intended message.

The metaphor allows us to bring out emotional elements; every time we use it, we give meaning to the objects we use.

The hand is the organ connected to mental development, and learning occurs through direct experience.

Exercise 5

Look around and choose an object that you see and that attracts your attention. Find the common elements between you and the object and build a metaphor/similarity, starting the sentence with the words "I am a/an (name of the object) …."

Some examples: "I am a pencil: I express thoughts and always leave a mark." "I am a blanket: I cover, I am warm, I can be used to hide and for comfort."

Object	Metaphor
	I am...

This writing exercise is also applicable to design because it allows us to find relationships and evoke concepts useful for the communication and purpose of the product.

Repeating these exercises several times will be useful for becoming familiar with these concepts, training our imagination, and developing our creativity. Our creativity will allow us to finalize the contents.

***Fernando*, napkin holder-place card, Progetti (1995)—design by Irene Guerrieri**

The little man *Fernando*, which was my first project put into production in 1995 by the company Progetti of Carate Brianza, can be considered a metaphor. Its concept is contained within its shape and its soft and rubbery material. It was born from the relationship I wanted to establish between the material, a synthetic rubber commonly called EVA, and the graphic of the little man frequently displayed outside bar and restaurant toilets. Attentive to graphic cues, I wanted to capture in that stylized figure, male and female, the necessary characteristics for the marriage with a material, of which I kept some scraps in my studio. I redesigned the graphics to suit my needs and began cutting out the shape from a sheet of expanded rubber I had available. I corrected the silhouette several times until I reached the optimal dimensions and characteristics. Thus was born *Fernando*, a flat rubber man who can assume various positions, thus performing different tasks, as every human being does every day. The name *Fernando* was the name of my doorman when I lived in downtown Rome; the daily routine of his job inspired the name of my project. Every day I passed through my gate several times and always found him there, vigilant, tireless, at the service of the condominiums, which were then inhabited by my relatives. *Fernando* was the first of a long series of products I designed and planned for different brands.

Now, let's try a small exercise in storytelling. The designer's manual mentions writing techniques for a reason: we convey ideas through writing, and the same must be done in designing objects, whether toys or everyday items. We can therefore consider the designer a bit like a writer of images. Words, like objects, are atoms, small portions of the world, to be put in relation to each other: the more words we have, the wider our world becomes.

The next exercise involves choosing seven objects with which we have to build a small story. For example, we choose apple, well, water, ball, child, tree, leaf. The story could go like this: an apple falls from the tree and bounces, ending up in the well, splashing out the water that wets a child playing ball with a leaf in his mouth. But with the same objects we can also say: a ball hits an apple that falls from the tree together with a leaf on a child's head who is drawing water from the well.

There are many storytelling games on the market, internationally known as *lacing games*, which stimulate the child to find relationships between different objects, building a short story. The proposed activity is to thread the individual elements of the game with a lace, respecting the sequence or the different sequences proposed. Each element can initially be considered individually and then become part of a narrative to be placed appropriately.

Lacing game

Exercise 6

Choose seven objects and write them in the appropriate boxes. In the row below, write the numbers from 1 to 7 following the order in which the objects appear in your story.

Object							
1st Succession							
2nd Succession							
3rd Succession							

You can change the sequence many times, inventing new and completely different stories.

The game lies precisely in the versatility and transformation of an idea, starting with seven fixed points.

If you think about it, this activity is typical of all construction games, which are characterized by some fixed elements, mountable and dismountable to obtain different results, often with a sequence in the assembly necessary for the final success.

Among the numerous examples on the market, some very structured, there is a game of extreme productive simplicity but of great value on the level of playful experience, aimed at the development of imagination, which explains this concept very well: *Tangram.*

Tangram is an ancient game, also known as "The seven stones of wisdom," which comes from ancient China. It is a tablet divided into seven parts (called *tan*) initially arranged to form a square; each part is a small object corresponding to a geometric figure (a small square, a parallelogram, two small triangles, a medium triangle, and two large triangles). With the seven elements present, you can compose infinite representations.

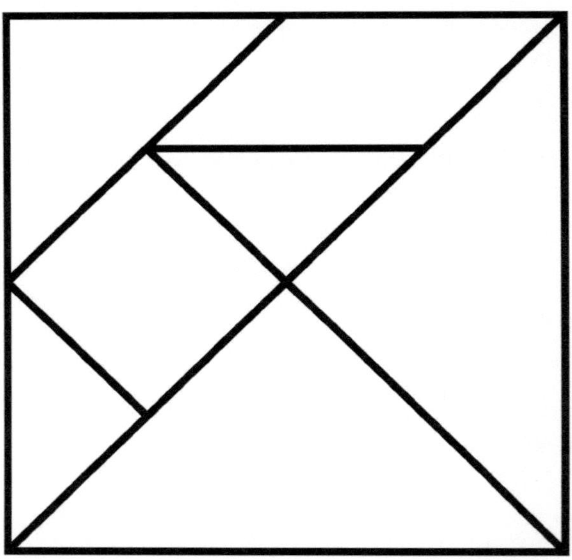

Tangram

The legend about the game's origin tells of a monk who gave his disciple a porcelain square and a brush, telling him to travel and paint on the porcelain the beauties he would encounter on his journey. The disciple dropped the porcelain square, which then broke into seven pieces. In trying to reassemble the square, he formed intriguing figures. He understood from this that he no longer needed to paint on the tablet, because he could represent the beauties of the world by reconstructing them in an infinite number of ways. One of the first books that talks about this legend dates back to 1817, and this game is called *Tablet of Truth.* In China it became an object of worship, and even Napoleon Bonaparte became a passionate player of *Tangram* during his exile on the island of Saint Helena. *Tangram,* in short, is a timeless game. Over the centuries, thousands and thousands of different figures have been invented with those seven pieces.

The aim of the game is to form meaningful figures. The rules are quite simple: use all seven pieces to compose the figures, and do not overlap any.

The game's main focus is the relationships between the seven elements and between each element and the figure parts you want to represent.

Figures made playing *Tangram*

An interesting contribution about relationships between objects and their succession in a narrative is given by the interactive books with holes, of which the publishing house La Coccinella is a pioneer. The idea of a interactive book had not been looked into prior to that, except for some trials by Bruno Munari.

In 1977, the publishing house La Coccinella was established, with the common goal of the four founding partners—Domenico Caputo, Giuliana Crespi, Loredana Farina, and Giorgio Vanetti—to create innovative projects capable of educating through play. As a result, the interactive playful book was created, and a particular series known as *Books with Holes* was developed. *Brucoverde*, one of the first books in the series, was released in 1977 and is still available today.

***Brucoverde*, La Coccinella (1977)—Loredana Farina and Giorgio Vanetti (authors), Giorgio Vanetti (illustrator)**

The cardboard interactive books include stiff pages, lots of holes for your hands to explore inside and find new figures, tiny windows to open, and tabs to pull. The reading of interactive books should involve all five senses. Exploring a hole therein, you first imagine, and then discover the next page, with another figure and another surprise.

Here, too, we are in the presence of objects put in succession, linked by relationships made of shapes, colors, and text, united by the pages that hold them together.

Brucoverde was exhibited at the Triennale di Milano, on the occasion of the "Giro Giro tondo—Design for Children" in 2017, as a book object part of the history of Italian design.

In this chapter we have learned the importance of the relationship between different objects when talking about design and how, through the connection between different concepts, something extraordinary and surprising can sometimes be born, capable of exciting and stimulating new imaginative mechanisms.

Knowing that one thing can also be another thing is a type of knowledge linked to mutation.

Mutation is the only constant of reality, and reality is made of relationships.

4

Decomposition

The Interesting Exploration of an Object Lies in Its Decomposition

Abstract The first experience every child has with an object is in its exploration; that means also its decomposition or splitting. A creative person is skilled at disassembling and reassembling an object by broadening his knowledge to various possibilities. A toy can help a child develop creativity. Through the toy, children can experiment with the rules of visual language.

We know that everything a child learns will remain in his memory and contribute to shaping his personality, and playful experiences will help him relate to reality. Therefore, toy designers need to remember to give children objects that will aid in their conceptual understanding and foster innovative interpersonal relationships. Exploration is the initial experience with every object. Exploring a toy object for a child involves getting to know it, understanding how to use it, and becoming familiar with it. From this it is easy to deduce that the interesting exploration of an object also involves its decomposition. A creative person can disassemble and reassemble an object, expanding his knowledge to the various possibilities of the case. A person who is minimally creative, or not creative at all, will experience greater difficulty in finding solutions and will always need external help in solving problems.

Through the toy, you can help a child develop creativity. How can we assist him with this?

Naturally, there is more than one response to the issue, but in this chapter we will attempt to understand the magic of "decomposition" and the freeing potential of this activity.

I. Guerrieri, *Toy Design*, https://doi.org/10.1007/978-3-031-98872-1_4

Let's explore games through which children are able to understand and experiment with new rules of visual language.

Exercise 7

Take sheets of paper and tear them by hand, producing different shapes, all irregular: squares, triangles, ovals, diamonds, and any strange shape that can be obtained by tearing the paper without rules. You will end up with both large and tiny pieces of paper which may be drawn on using pencils, colored crayons, or other tools, depending on what the shape indicates.

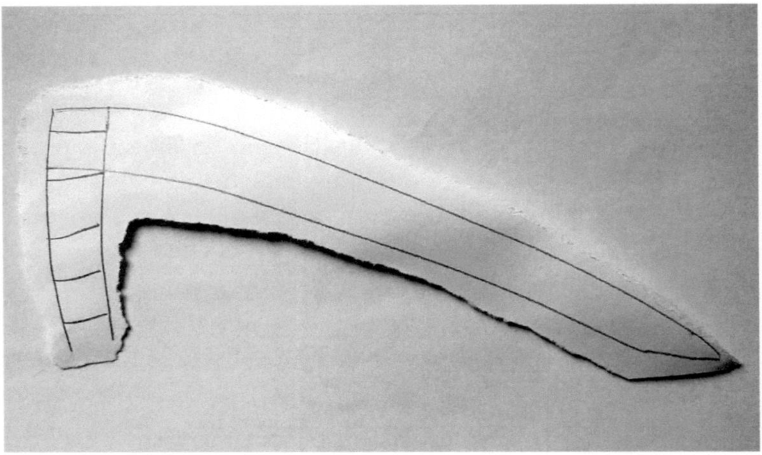

A piece of paper that looks like a slide

This exercise stimulates the production of significantly variant figures.

A single sheet is less stimulating than its division into many shapes, and each shape can be viewed differently by individuals.

You can repeat the exercise several times and share it with other people. The results will be surprising, always different, and will often cause healthy and carefree laughter. The goal has been achieved, as the game presupposes fun.

A similar activity can be proposed in the field of three dimensions, using materials that are easy to shred. We could have fun doing the experiment with a piece of foam rubber or tearing a piece of Parmesan cheese into pieces with our hands; we would undoubtedly get very irregular shapes, which, if we are able to look with a creative eye, will send us back to images from our memory. We might see a face, or a whole human figure, sometimes an animal, or a sailboat. Each random shape recalls images from our past and lets us experience visual language.

Staying on three dimensions, let's try to consider fruits divided into several elements. Take, for example, an orange, which has slices pre-packaged by Mother Nature that are easily separable, making the object easily decomposable in all its parts. With the slices of an orange, we can form a flower, in different compositional modes, or the wheel of a bicycle; if we then consider the single slice, this immediately brings us back to the image of a smile or of a bridge; two slices instead to a mouth that can open and close. And so, with a simple orange, we can play and have fun decomposing it and rebuilding objects belonging to different categories, activating connections with the images of our knowledge. Before being eaten and therefore fulfilling the function for which Mother Nature gave it to us, an orange is like a toy, capable of stimulating creativity in those who allow themselves to be stimulated.

An orange with its segments

Even an apple or a kiwi can be broken down, with the help of a knife, into sub-elements of different shapes that remind us of images well present in our memory. We can use the segments of an apple to build a flower; if instead we take the single segment, we immediately think of a boat. If we slice a kiwi, the round shapes thus obtained remind us of eyes. If we then combine the elements of several fruits, we can have fun building faces.

An apple and its segments

Two slices of kiwi

To continue with these experiments, we must exploit a channel, always active in children, which is that of curiosity.

Exercise 8

With our curiosity, we are going to investigate the shapes of nature: flowers, fruits, insects, and animals. In the first column, we indicate the shape's name, and in the second column, we describe the well-distinguishable internal element into which we can break down the initial shape. Let's make a brief list.

Whole shape	Single component element of the whole shape

In design, we talk about "morphology of components" when we refer to the study of the shape of the individual components of an object. This analysis is essential for industrial and serial productivity. We address the uniqueness of each individual component and the specific characteristics of the design to optimize productivity. Sometimes, some components belonging to an object can also be used for other objects. Other times, some components constitute real modules, which are repeated and juxtaposed to obtain different results. When a component becomes a module, the final object is versatile and modifiable.

Exercise 9

An interesting exercise can be to try to draw a two-dimensional geometric shape that can be defined as a "module." A module can be juxtaposed with a similar one in a unique and original way. Try to invent a two-dimensional module using the underlying grid. Then copy it and cut it out several times, and try to combine more modules together to obtain a unitary shape interesting for its aesthetics or its function. This is the reverse operation of "decomposition"; it is called "composition." To better understand the first, we must also experience the second.

Draw your module using this grid.

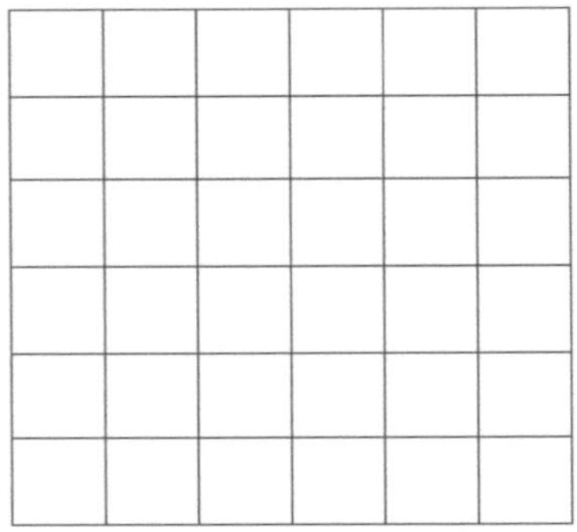

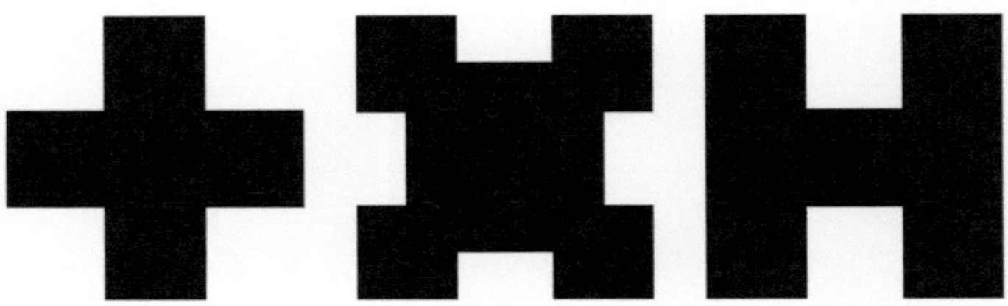

Some examples of modules

Even in writing and music we can talk about "decomposition" and "composition"; the modules are constituted in the first case by letters and in the second case by musical notes. A word can be broken down into parts of a word that, when reassembled in a different way, can take on different meanings:

TRA	IN
IN	TRA

R	O	M	E
M	O	R	E

Decomposition and recomposition of words

This means that even writing and music can be taken as a game, to be decomposed and recomposed creatively: everything lies in identifying what the module is. We commonly talk about "musical composition" and "poetic composition," but the fun game lies in their "decomposition" and recomposition.

We must learn to observe nature if we want to understand the structure and harmony, as well as the brilliant idea of many things that surround us: a beehive is formed by the linear repetition of many small hexagonal cells, while an orange is formed by the circular repetition of its segments.

Beehive

Slice of orange

Any subject can be explored by breaking it down into its components. A human body can be broken down into its main parts—head, arms, legs, and chest—and then be reassembled in an unnatural way, with legs in place of arms and arms in place of legs, or with the head in place of legs and legs in place of the head, giving rise to the fantastic theme.

In the *Zookids* line, designed by me for Sevi/Trudi, I created a game of interlocking between three animal figures—a zebra, an elephant, and a lion—where the decomposition of each of them into four distinct parts of their body allows the interchangeability of the elements, giving rise to fantastic animals: an elephant with the lion's mane, a zebra with the head of an elephant, a lion with elephant legs, etc. With children, you can laugh at such absurdities and recognize the error that gave rise to the fantastic theme born with the decomposition.

***Interlocking*, line *Zookids*, Sevi/Trudi (2002)—design by Irene Guerrieri**

Each element retains its original meaning if taken individually and assumes another meaning if considered in its entirety.

Longer & Shorter is an educational card game with illustrations, which I created for the company Headu in 2019, that utilizes the theme of decomposition. The figures, once broken down into multiple parts, are reconstructed shorter or longer depending on whether an additional card, called a *Joker* card, can be used or not. You can play different games with this deck, enabling you to choose whether to use the extra element based on the game's goal.

Longer & Shorter, Flash cards, **Headu (2019)—design by Irene Guerrieri**

I cannot fail to mention here also a line of products, which I created in 2013, *Rebuild the Monument*, where the proposal lies in the decomposition and reassembly of the identifying monuments of the most important cities in the world. All parts of the same monument, or square, are contained within a soft suitcase that bears the name of the city to which it belongs on the outside. The soft, slightly padded elements with Velcro are easily attachable to the base fabric, thanks to the particular structure of the fabric: a tricot. A nice travel souvenir to remember a place visited, where memory and compositional ability come into play to reconstruct the visual image present in one's memory. From the suitcase, pieces are extracted, each clearly distinguishable: column, facade, dome, colonnade, etc. The child is invited to grasp their meaning and the relationship that exists between them to bring the monument to life.

***St. Peter—Rome*, Rebuild the monument (2013)—design by Irene Guerrieri**

If we consider the basic geometric figures, an interesting three-dimensional example for decomposition is the cube.

A cube broken down into nine parts can give rise to a fun construction game. If the sub-elements we obtain (prisms and cubes), which are derived from the large cube, become the representative elements of the nativity scene, then the "decomposition" refers to narrating a story: the story of all stories. A playful way to experience the nativity scene as decoration and story, where the individual characters appear one at a time in the scene, talking about themselves and always placing themselves in different ways, stimulating language and imagination. The *Nativity Cubotto*—this is the name I gave it—was a self-produced project for Christmas 2016, and then reproduced in subsequent years.

Nativity Cubotto **(2016)—design by Irene Guerrieri**

I am particularly interested in exploring the decomposition possibilities of the cube object. We can indeed decompose a cube in infinite ways, opening it and making it become a two-dimensional object, examining the flat geometric figures that compose it, but also considering its three-dimensional volume and sectioning it in different ways. The elements obtained stimulate new constructions and different conceptual ideas.

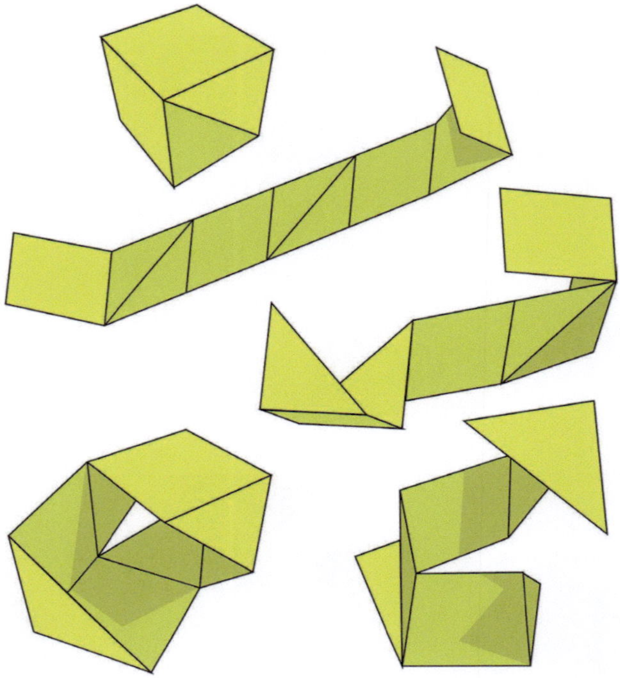

Open cube

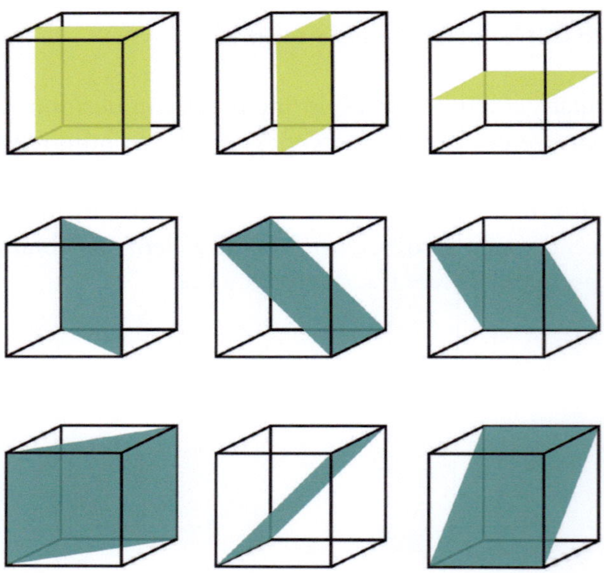

Cube sectioned in orthogonal and oblique mode

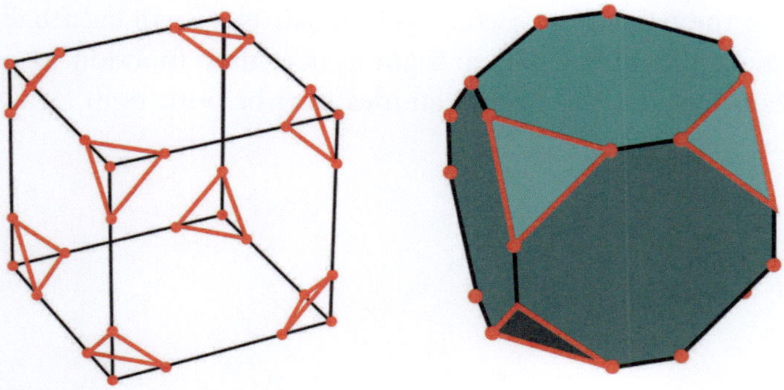

Cube deprived of edges

In the same way, we can explore the sphere and the pyramid.

Exercise 10

Decompose a cube, a sphere, and a pyramid, and write something that you can imagine with the decomposition made. Keep in mind the example of the nativity cube, but imagine something else.

Volume	Type of decomposition	What you imagine with the decomposition made
Cube		
Sphere		
Pyramid		

Sketching a budding idea freehand is a very important exercise that I always recommend. Sketching and jotting down your insights is a bit like laying out the plot of a book.

The first notes of an idea constitute its conception, which, if nurtured, will take shape and life. A notebook dedicated specifically to this type of exercise allows you to preserve the chronological path of an idea's birth from its beginning; page after page, we can retrace it at any time. A notebook is comparable to the gestation time of a pregnancy, and as such should be lived intensely.

Preserving the primordial sketches is crucial, as we will eventually need to revisit them to comprehend their significance; they function as our guides and sources of inspiration. Even if an idea can't be born right away, it can be revived later when it's ready.

5

Union and Separation

Every Concept is Impossible Without Its Opposite (Paul Klee)

Abstract Two distinct parts, joined together, become a surprise, an exciting experience, giving satisfaction when they come together correctly. Many games are based on this principle, like puzzles and many construction games. Through this kind of activity, you can learn many concepts.

We must consider how every idea gives rise to a binomial of ideas, and there is no concept without its opposite. Just as there is no "composition" without "decomposition," in the same way we cannot talk about "union" without "separation." The puzzle game is based on this principle.

I. Guerrieri, *Toy Design*, https://doi.org/10.1007/978-3-031-98872-1_5

The Space, Sassi Junior (2015)—design by Irene Guerrieri

Two distinct and separate parts joined together become a surprise, an exciting experience that gives satisfaction when they meet and correctly hook together. They correspond to the achievement of a result.

By always staying on the theme of the puzzle, we can find the same type of activity in other game concepts. An example is the book with removable pages designed for Sevi/Trudi, in which the image must be recomposed by correctly hooking the pieces.

Wooden booklet with joints, line _Zookids_, Sevi/Trudi (2002)—design by Irene Guerrieri

Through this activity, concepts can be learned. Each element of the book puzzle has a different illustration on both sides. The child must hook the pieces in the correct way to reconstruct the figure exactly, giving readability and sequence to the wooden pages that make up the book.

The following image introduces the concept of union-separation in an educational wooden game that teaches numbers from one to ten. From the union and separation of the fish with its tail, the child is invited to hook the pieces correctly, matching the illustrated number to the number written in figures, and obtaining at the same time the correct figure of the fish. In this proposal, the color provides a little help, adding extra skill to the game, given by the recognition of colors.

***Fish*, Innovation in a Box (2013)—design by Adittee Naargolkar, Irene Guerrieri**

The Chinese have handed down the famous symbol yin-yang, a unit in the shape of a disk formed by two equal and opposite elements: one white and one black. Each element is inverted in relation to the other.

These two elements represent the balance of life. The concepts of yin (black) and yang (white) originate from ancient Chinese philosophy and refer to the day and the night, reflecting also in every aspect of nature and in every element present in it.

A part can only be understood in relation to the whole.

Chinese symbol Yin-Yang

Exercise 11

Make a list of pairs of objects that are opposites or reverses, then draw these objects in your notebook.

First part	Second part

All balancing elements are exactly the opposite of what caused the imbalance and tend to fill it.

In many children's toys we see this concept materialize, not only in the pieces of the puzzle already mentioned, but also in every type of proposal where the child is called to find the missing element to meet balance and completion criteria.

In *16 Animals,* designed by Enzo Mari, which has been included in the Danese Milano catalog since 1957, the silhouettes of the animals become perfectly harmonized interlocking elements.

***16 Animals*, Danese Milano (1957)—design by Enzo Mari**

In *Animals at Zoo* by Innovation in a Box, the completion criterion is provided by the proposal to match the animal to its own texture of fur. The interlocking only works if the match is correct.

***Animals at zoo*, Innovation in a Box (2013)—design by Adittee Naargolkar, Irene Guerrieri**

A game where the completion activity required of the child also involves the use of paper and pencil is that of the game *Faces*, designed for Lisciani Giochi in 2012. The game involves using a deck of illustrated cards in various ways to reassemble the faces of ten animals. In the absence of one or more cards, children are invited to complete the figure by placing those in their possession on a sheet and drawing the missing elements in pencil. The completion game, with or without drawing, is very satisfying and is the reason why many games are set on this principle. Combining cards in the correct way, to obtain the right figures, is an activity that satisfies the criterion of the union while at the same time leaving the possibility of separation in order to start playing again.

Faces, Lisciani Giochi (2012)—design by Irene Guerrieri

6

Transformations

Deformation and Error Stimulate Creative Ideas

Abstract Considering an object in all its possible transformations is an activity that children really like. From a starting idea, coming from memory, by applying a series of transformations linked to other experiences and other ideas, something new is created. Basically a transformation is applied to the initial idea. Children love to do this, and the designer has to go back to being a little kid to apply this tool.

The theme of transformation is dear to all creatives. Observing an object and then considering all its possible transformations is an activity that children love; they enjoy recognizing new appearances and attribute new meanings to every common thing they find.

In this context, I wish to highlight the phenomenon of shadows cast on walls, which has consistently captivated youngsters across cultures and epochs due to its ability to create and elicit distortions, so altering and amplifying simple things or minor body parts. A hand transforms into the mouth of a wolf, into a rabbit, into a mouse, or into a crocodile if, by assuming a particular position and using a light, we project its shadow onto a surface. You can add to the scene some small details like a ring, a ribbon, or an elastic band, which, because of the projection, deforms and enlarges, so what we have illuminated transforms into new appearances, visible only through the shadow created.

I. Guerrieri, *Toy Design*, https://doi.org/10.1007/978-3-031-98872-1_6

Light and shadow, a perfect pair, give life to new creative ideas thanks to transformation.

I have created numerous whimsical items derived from a personally significant concept, selected from my experiential repertoire, guided by my instincts and emotions, to which I subsequently applied various transformations informed by other experiences. From a primordial idea within my memory, combined with other ideas, I allowed my imagination to flourish and created something new that did not exist before. I essentially applied transformations to my starting idea.

In 2003, in collaboration with the publishing house La Coccinella, I designed a series titled "Funny Books," containing two titles: *The Cheese Ghost* and *The Frog Bastiana*. In the first title, I started with the image of the ghost, thinking of a sheet with two holes in place of the eyes. This image, which is dear to me, served to bring the interactive book to life through a process of transformation: behind a sheet with two holes, a child hides; the silhouette of the sheet transforms into the silhouette of nothing on the next page because the ghost does not actually exist and thus corresponds to a hole in the wall. Subsequently, other images stemming from this initial idea contributed to the conception of the interactive book, whose rhyming story was written later.

The same goes for *The Frog Bastiana*, whose starting image was instead the facial silhouette of the frog, from which other images were born in succession due to the transformations that I chose to apply. From one page to another, an image transforms and revives into new appearances. The game lies in the surprise and in the satisfaction that comes from learning the process of transformation. In a interactive book, the product idea comes first, then the text.

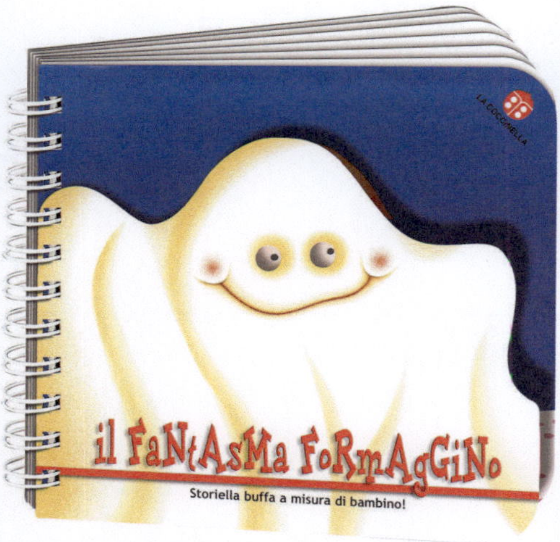

***The cheese ghost*, La Coccinella (2003)—design by Irene Guerrieri**

Exercise 12

Cut out rectangles of paper about 10 × 15 cm; the number will depend on the elements to be represented. A story is chosen, and its identifying elements are represented on the prepared rectangular cards. Each element must be drawn on a separate card. If we choose the story of Little Red Riding Hood, for example, six cards are enough for the following six elements to be represented: mother, Little Red Riding Hood, basket with snacks, wolf, grandmother, and hunter. Next, line up the cards in the correct sequence, and use scissors to transform each one. Through holes or cuts, we create communication between one card and the next.

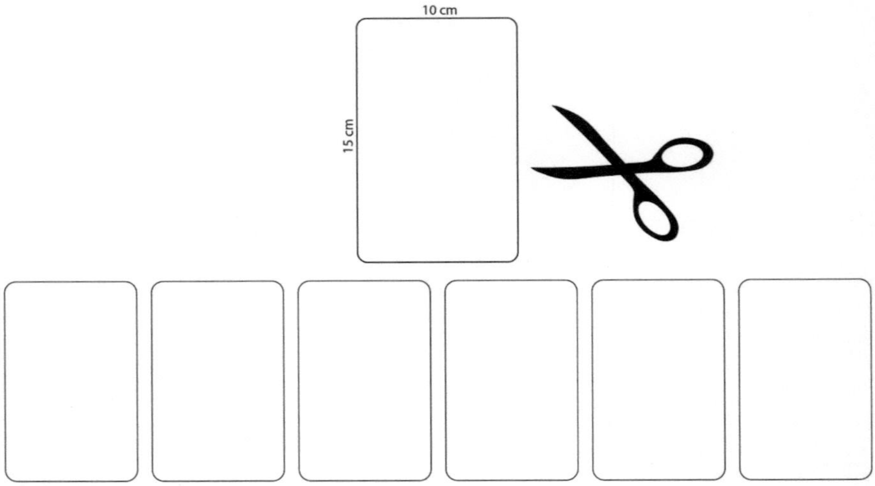

Cards

This exercise helps us understand how to activate transformations to connect one element to another when we are faced with a sequence of figures.

Nature itself shows us the extraordinary power of transformation in the seed that transforms into a flower and then into fruit.

SEED → FLOWER → FRUIT

To explain this process, the company Cocai Design has created some cute *leporellos* that illustrate with images and silhouettes the transition from flower to fruit.

***From flower to fruit*, Leporello, Cocai edu, Cocai design (2019)**

I have often had the opportunity to design educational games aimed at learning about life cycles. The butterfly's life cycle, including the caterpillar and chrysalis, is the most requested kindergarten theme. The butterfly is a wonderful insect; a symbol of freedom. The process of transformation through which it develops and transforms into an insect is called metamorphosis. The butterfly lays eggs, and from each egg, a larva hatches that grows and changes its skin several times, eventually becoming a voracious caterpillar that consumes various types of leaves and vegetables. The caterpillar eventually stops eating, builds a protective mantle, the cocoon, and falls into a deep sleep, transforming into a chrysalis. After a while, it becomes a butterfly that breaks the protective mantle and emerges from the shell. It is no longer as voracious as it was when it was a caterpillar; instead, it utilizes its lengthy proboscis to extract nectar from flowers.

This wonderful story is narrated in a toy using a spinning wooden circle in which the different pieces must find the correct spot.

***Life cycle of butterfly*, Innovation in a box (2013)—design by Adittee Naargolkar, illustrated by Irene Guerrieri**

Another fitting example is the small cutlery set called Traffic by Fratelli Guzzini, which transforms itself into means of transport, involving a change in both meaning and form while maintaining its function.

The scaled-down means of transport are used to transport food.

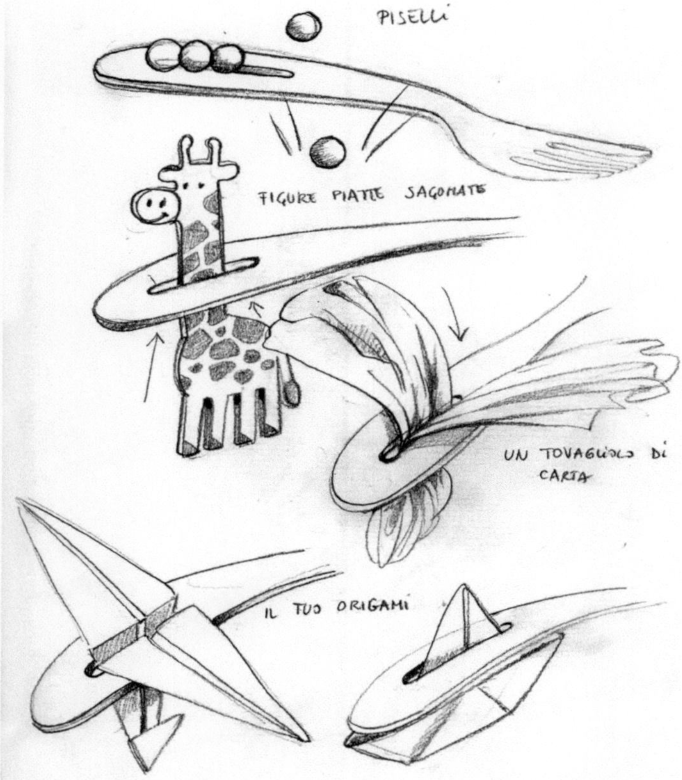

Sketches of small cutlery, line *Traffic*, Fratelli Guzzini (2017)—design by Irene Guerrieri

Small cutlery, line *Traffic* and *Play Park*, Fratelli Guzzini (2017 and 2019)—design by Irene Guerrieri

It can be great fun to represent the transformation of some figures when they coincide with a real deformation, and fun is a fundamental ingredient of any game.

The deformation of an object or a figure can arouse new emotions, spark new fantasies, and produce new ideas.

In *Zebra*, which I designed for Sevi/Trudi, as described in Chap. 3, the transformation applied to the animal as a pull toy made the object attractive and fun.

In *Longer and Shorter* by Headu, fun deformations can be arrived at by transforming the figures to be reconstructed with the cards, choosing whether or not to use the Joker card.

Longer and shorter **(preliminary studies),** ***Flashcards*****, Headu (2019)—design by Irene Guerrieri**

A small ship can become a transatlantic vessel, and a utility car a limousine.

Children love deformations. Creating deformations is not only fun, but also a productive and functional game because it helps to explore, stimulates freedom, encourages nonconformity, and, above all, makes you laugh.

The initial image is elevated and rendered significant through deformation.

I designed a wooden stork with long, soft legs for Sevi/Trudi. Pulling a red cord opens the mouth and causes the legs to move. The transformation applied to the animal made the object fun and playable.

Stork**, line** ***Zookids*****, Trudi/Sevi (2002)—design by Irene Guerrieri**

Exercise 13

Fill in the two parallel columns. In one, write the object's name or image, and in the other, its deformation.

Object	Deformation
Giraffe	Super long
Stork	Lying down

Pairing can often result in a fun and stimulating idea.

The Lying Stork was an idea for an interactive book published by Gribaudo. No one has ever seen a stork lying down, yet its unusual and exaggeratedly elongated position gave birth to an idea: a book with detachable pages to reconstruct to obtain the image of a long, lying stork.

Part of the same series is *The Spotted Banana,* where a long banana is presented in pieces. Each page of the interactive book corresponds to a piece of banana and a portion of text. You can lengthen or shorten the banana as you wish, adding or removing pages and lengthening or shortening the story at will.

The Spotted Banana is so named because it has brown spots on its peel. If you look closely at those spots, you can recognize the semblance of a cat, a shark, a kite, or whatever you want. On the backs of the pages, the banana is presented only as a white silhouette to color, where you can draw the spots you want, inventing and fantasizing a new story.

The interactive book allows transformation, and the reader becomes the author himself.

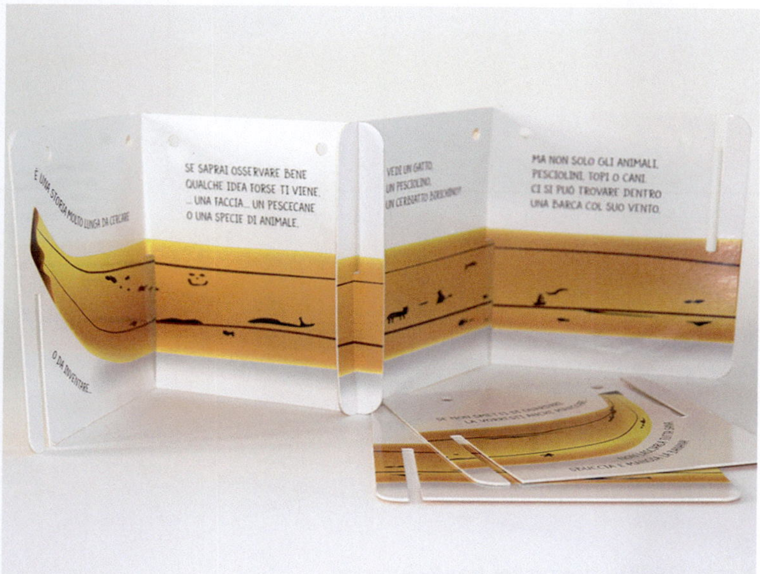

***The spotted banana*, Build your own book, Gribaudo (2017)—design by Irene Guerrieri**

In the tale of Pinocchio, we witness two transformation processes: that of the piece of wood becoming a puppet and then a real boy, and that of the nose that lengthens and shortens.

Around the transformation of Pinocchio's nose, a game was born, which I called *PinOOOcchio*, emphasizing in the name the attention that the player must pay to the lengthening of the nose.

Below is the version with cards, where you can achieve a transformation depending on the placement and availability of the nose cards in your possession.

***PinOOOcchio* (2008)—design by Irene Guerrieri**

Another way to apply transformation is to give voice to the creative error.

As Rodari teaches us, the spelling error in writing, if well considered, can give rise to comic stories with meaning: a child who writes "the needle of Garda" suggests the story of a needle so important that it's even marked on the geographical map. (In Italian, if you write "L'ago" instead "Lago", you traslate "The needle" instead "The Lake")

In design, we can find visual errors that suggest new ideas and introduce new concepts.

In every error, the possibility of a creative idea is born. This is true for both writing and design.

From a childish error, therefore, a new creative and stimulating concept can be born.

Laughing at humorous, unusual representations that one does not see in reality is a creative action.

For the Thun company in 2005, I designed the pushable penguin with soft rubber legs applied to wooden wheels, which make noise when turning. Each wheel is equipped with two soft legs, for a total of four legs. I wanted to emphasize the penguin's legs in a deliberately exaggerated manner to make the toy more attractive and fun.

***Pushable Pinghi*, children's line, Thun (2005)—design by Irene Guerrieri**

My 2013 game *Faces in My Food* provides another example of applying the creative error. Here, the child's creativity and inventiveness bring the creative error to life by conceiving food as part of a transforming puzzle.

With food, you can play and invent strange faces.

Faces in my food (2013)—design by Irene Guerrieri

This concept deviates from reality, leading us to view food as a source of play and fun. A new meaning is given to food, and the focus is on the usefulness of its shapes and colors. This game can be played at the table with real food, making mealtime fun. In the 1500s, Arcimboldo created human heads with fruit, fish, branches, roots, and leaves. Each part loses its original meaning and assumes another as a result of the way it is used.

Imagination manifests itself in the transformation of some situations. You can transform an object by applying its opposite to it: a super-fast turtle becomes a fun toy. There are transformations in the change of place, color, size, weight, and function. Also, the fusion of different elements into a single body can give rise to new figures. A beautiful example in this sense is the *Universal Bestiary* by Professor Revillod, published by Logos, where elements of different animals, put together with a certain compositional skill, give rise to monsters with very original names. In the field of furniture, some pieces feature legs shaped like lions' feet, while others have legs resembling those of different animals.

Exercise 14

Fill in the columns. In the first, write the name of the object you want to represent; in the others, its transformation of place, color, size, weight, and function.

Object	Size trasformation	Color trasformation	Place trasformation	Function trasformation
A loaf of bread	small	Purple	In the sea	It was a boat

In the art of origami, the transformation occurs from a two-dimensional form to a three-dimensional one. Origami is an ancient pastime, currently taught in Japanese elementary schools, which consists of creating figures by folding and refolding colored paper sheets.

In the transformation from the two-dimensional sheet to the three-dimensional object, there is a great surprise and a sought-after creative act.

Origami

Japanese children, educated from an early age in this type of art, easily understand the transformation from two-dimensional to three-dimensional.

In *My Christmas Tree*, published by La Coccinella in 2015, this type of transformation is present. When opened, the shaped pages compose a Christmas tree, full of windows and openable parts that transform its appearance: from a two-dimensional figure, the tree becomes three-dimensional.

***My Christmas tree*, La Coccinella (2015)—design by Irene Guerrieri**

Discovering that one thing can become another is a type of knowledge linked to transformation. Transformation is the only constant of reality.

7

Mathematics in Games

You Can Start With a Logical Structure for the Invention of a Game

Abstract We can talk about the choice of mathematical language in games as an indispensable training proposal for the child. Children tend by nature to group and order objects and materials according to different criteria; they are interested in taking measurements, they enjoy positioning objects in space, and they enjoy following paths. Through the game it is possible to acquire temporal, spatial, and measurement concepts.

The construction of mathematical thinking is a progressive process toward which the child tends and through which new concepts, skills, competencies, and attitudes are acquired, intertwined, consolidated, and developed repeatedly; it is a process that requires a gradual acquisition of its language. We can talk about the choice of mathematical language in games as an indispensable educational proposal for the child. The purpose of designing a toy should always be to provide early childhood play experiences that promote a positive, enjoyable, and stimulating image of mathematics. The goal is for children to acquire useful skills in this field that will aid in the development of their own thinking, language expression, judgment, reasoning, and social relationships.

Mathematics is the basis for everything. In learning mathematical concepts, the child feels satisfaction and experiences self-esteem.

Children instinctively tend to group and order objects and materials according to different criteria. When given the right kind of stimulation, kids are able to distinguish between various qualities by comparing and assessing various quantities; they are interested in performing measurements using

I. Guerrieri, *Toy Design*, https://doi.org/10.1007/978-3-031-98872-1_7

tools available to them. They enjoy positioning objects and people in space and following paths based on verbal indications.

Through play, various concepts such as forward/backward, above/below, and right/left, as well as measurement concepts, are acquired.

In many toys, the logical structure that helps children open their minds to set-related concepts is clearly visible; in other cases, activities related to numbering and sequencing are clearly visible. Classifications of objects, belonging of shapes and colors to certain categories, and numbering are activities required in many toys, and they help the child to achieve results and derive satisfaction while having fun.

Children need to understand with their heads, but also using their hands and their imagination; logical concepts that they can apply in various fields.

We can start with logical reasoning to invent a toy or go the opposite way, starting from a fantastic idea to arrive at the conception of a toy to be used to learn logical concepts. In both cases, the logical structure influences our design.

Designing a toy with a focus on its logical structure involves creating relationships among the elements that make it up, based on logical concepts. Some toys are based on the relationships big–small, high–low, and little–much. It is stimulating for a child to play with the relationships of opposites. Some of these activities are based on measurement or on the ability to count. Others introduce notions of set theory or simple mathematical operations. All these activities represent the first achievements related to problem-solving.

In the game *Queen Little Queen*, to the question "How many steps do I have to take to get to your beautiful castle?", the child is invited to take a certain number of elephant steps, or lion steps, ant steps, shrimp steps, etc. The space is measured according to different units of measure, corresponding to the different types of animal steps called into play. Children interpret with their own imagination how to perform that step, identifying with the requested animal. From this game, you can learn that there are many ways of measuring; it depends on the unit of measure called into question.

The foundation of every scientific activity is measurement. From the game of *Queen Little Queen*, other fun mathematical exercises and games are inspired.

Measure Me Right is a game that was designed specifically to encourage a playful measurement activity, where the unit of measure is a small foot and the objects to be measured and compared are the animals of the zoo.

In the invention of games with mathematical content, we cannot ignore numbers: we normally look for relationships between these and the objects that satisfy our fantastic idea.

Measure me right, animals at zoo, **Innovation in a box (2013)—design by Adittee Naargolkar, Irene Guerrieri**

In *Numbers in my Garden*, this goal was achieved by creating a colorful blooming garden, where each number corresponds to a certain flower identified by its color.

Each flower is broken down into three recognizable parts, and in each of its parts—stem, leaves, flower—the number is represented in three different ways: digit, word, and dot illustration.

The activity of breaking down and reassembling the small flower garden on a leaf-shaped base allows the child to learn many things related to numbers, colors, flowers, and their parts.

***Numbers in my garden*, Innovation in a box (2013)—design by Adittee Naargolkar, Irene Guerrieri**

In *Butterfly Math Game*, the child is invited to decorate the wings of a butterfly toy using colored geometric elements, solving the first concepts related to basic mathematical operations.

***Butterfly math game*, Innovation in a box (2013)—design by Adittee Naargolkar, Irene Guerrieri**

The *Tombola* game also offers a playful activity related to numbers and counting. The *Tombola* designed for the *Animals at Zoo* line is an illustrated bingo in which the child is invited to count the figures and establish the correspondence with the number drawn.

***Tombola, animals at zoo*, Innovation in a box (2013)—design by Adittee Naargolkar, Irene Guerrieri**

There are games that teach us to classify, to build possible sets, and to exclude impossible sets of animals and objects. Both the children's emotions and their perspective on reality will be shaped by this merging of imagination and thinking. The exploration of sets is an exciting adventure for the child.

***Professions*, Innovation in a box (2013)—design by Adittee Naargolkar, Irene Guerrieri**

The child performs a fundamental operation of logical order when choosing an object to associate with another. The product is emotional reinforcement. The child invents beautiful stories by manipulating logical blocks. From a blue triangle looking for its place among the red squares, the "Ugly Duckling" is imagined trying to position itself among the white swans. There are possible sets and impossible sets, and the child builds the foundations of reality through reasoning.

Additionally, there are games that encourage children to go on adventures in search of the smallest or largest objects. Stackable towers are a well-known example of this type of toy; to build one, you stack pieces from biggest to smallest, and to dismantle one, you thread all the elements back into each other in reverse order. If we then associate numbers with each element of the tower, we can learn to count from 1 to 10 and from 10 to 1, clearly understanding that the largest number is associated with the largest object and the smallest with the smallest object.

My First Numbers, which I designed and illustrated for Sassi Junior in 2015, consists of a suitcase containing ten colored cubes to stack or insert into each other, accompanied by an illustrated booklet to learn to count. The tower is over a meter high, and on the faces of the cubes there are cute figures to count, corresponding to the element's number.

The pleasure of knowing and the desire to assert one's knowledge strengthen the child's sense of security. Educational games focused on learning and knowledge are particularly stimulating.

The riddle in this context plays a key role. Through it, we transform a difficulty into a solvable problem. The riddle is a form on which we can work to solve a problem. We can compare it to a method or an effective and original model for problem-solving.

***My first numbers, Eco cubotti*, Sassi Junior (2015)—design by Irene Guerrieri**

This type of activity is serious. The riddle in the game comes in various forms. The most loved by children is the classic "guessing game," in which one player thinks of an object, a person, or an animal, and the other player attempts to guess what it is by asking a series of questions, accumulating answers (data), and gradually narrowing the field of possibility, ultimately arriving at the answer. The method adopted to gradually narrow the possibilities is classification and grouping of data obtained from the received responses.

There are many games that contain this type of activity; a famous one is *Guess Who?*, known in different versions around the world, created by Ora and Theo Coster.

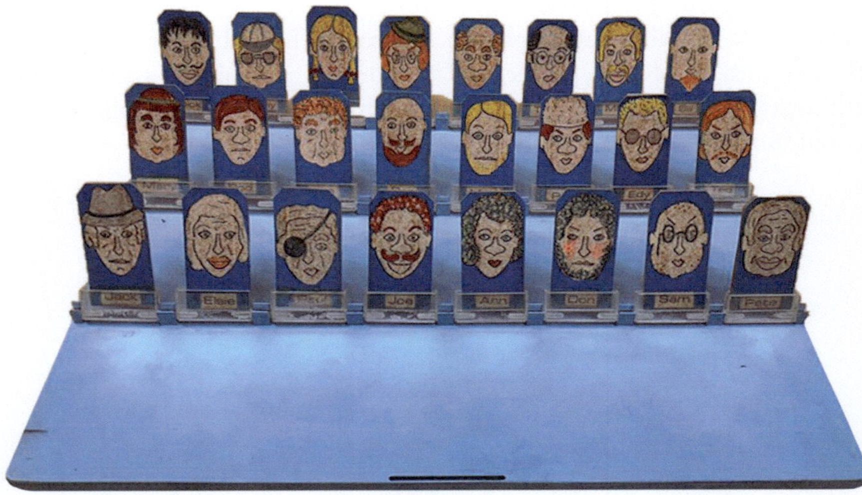

***Guess who?* (prototype 1978)—developed by Ora and Theo Coster, marketed by Hasbro Gaming, photo by B. Coster**

Even card games like *The Opposites*, produced by Arbos in 2016, involve remarkably similar activities. The box contains cards that illustrate representative subjects of a certain concept and match them to their opposites.

The differing game modes all focus on the classification of data. Background colors help younger children to match pairs.

***The opposites*, Arbos (2016)—design by Irene Guerrieri**

Even the games of *Memory* and *Domino* are mathematical games, where association and grouping skills come into play, allowing the child to acquire different concepts.

A great invention was Rubik's Cube, conceived by the Hungarian architecture professor and sculptor Ernő Rubik in 1974 and considered by many the best-selling toy in history. To solve the puzzle, each face must show only one color.

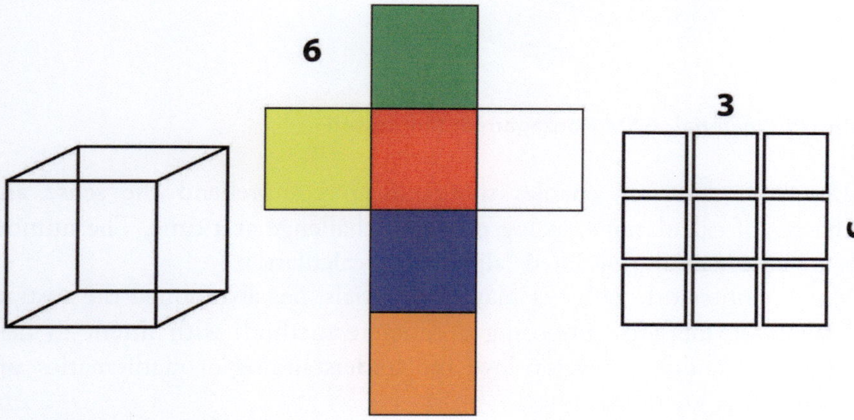

Logical structure of the cube

In Montessori schools, mathematics is the subject that children like the most: "The more complicated the topic, the greater the child's enthusiasm" (Montessori, *Education for a New World*). Maria Montessori has given children the educational material necessary to see, touch, and manipulate the quantities, even large ones, and to construct "large numbers." Focusing on the child's potential, Maria Montessori has proposed mathematical materials that allow active discovery of laws and theorems that in other schools must be learned with effort, relying on memory. She understood that children's minds only learn things that resonate with their knowledge.

***Montessori material*, GAM Gonzagarredi Montessori**

Montessori Material enables the child to comprehend the sense and mechanics of calculations, facing only one challenge at a time. The number rods are an essential tool called "alphabet of calculation."

Maria Montessori, with her playful materials, has abandoned the abstract and symbolic method, proposing a concrete method, with intent to help thousands of children develop love and understanding of mathematics and to discover the joy of learning.

***Montessori material*, GAM Gonzagarredi Montessori**

8

Shapes and Colors in Fairy Tales

How to Extract the Concept of a Game from a Fairy Tale

The fairy tale is a medium–short narration centered on events and fantastic characters, which need to be imagined to be understood.

The fairy tale is a story that needs images, which are sometimes provided by the book and sometimes by the reader's imagination. Whether in the first or in the second case, reading and understanding concepts through images is a serious activity, which allows us to associate shapes and colors with the things we encounter in the narration and that we in turn use to express concepts. This type of activity linked to the fairy tale is also applicable to toy design, which considers the shape, space, and color to tell and transmit concepts, just like an illustration does, with the only difference being that it uses three dimensions.

Why this analogy? Because through images we learn concepts and recognize shapes and colors within a fairy tale. We can achieve the *concept* of an educational game.

Today, there are beautiful, illustrated books on the market, with refined and high-quality images, that express emotions and feelings and help the child get into the *mood* of the story.

Line, shape, space, and color have great powers in conveying concepts. Understanding and helping others to understand is an activity that involves visual art; a toy fits well within this scope.

I. Guerrieri, *Toy Design*, https://doi.org/10.1007/978-3-031-98872-1_8

Let's see together where to start and how we can draw inspiration from a fairy tale to do this. Placing shapes and colors within a space involves crucial design concepts.

We consider fairy tales an inspiring source: our well of ideas from which to draw material and extrapolate concepts. As we did with the "images of memory" in Chap. 2, let's now immerse ourselves in the fairy tale, explore it, and discover its contents through shapes and colors capable of telling it.

The fairy tale becomes a dynamic game of shapes and colors; knowing how to read it means digging into the depths of our memory. Each shape and each color conveys an emotion or a concept.

The shapes that we can grasp are always inserted in a context. A red triangle in a fairy tale set at sea can refer to objects from our memory, such as the sail of a boat, a volcanic island, or the fin of a shark.

If we want to represent the fairy tale of *Little Red Riding Hood,* the red triangle can be representative of the character.

Maybe we need two, one small and one larger.

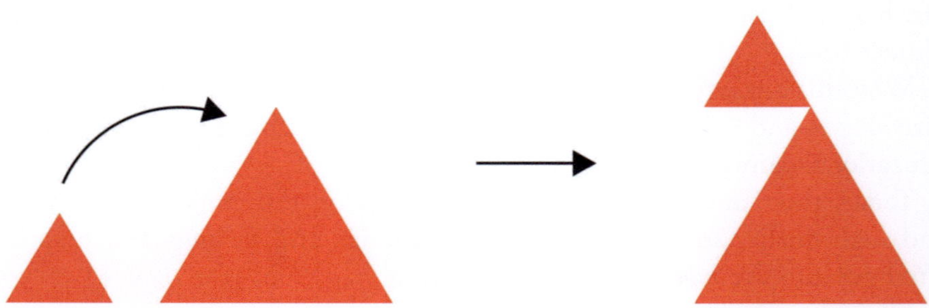

Then draw a circle for the face, and the game is complete.

We can also add two more small ones for the feet.

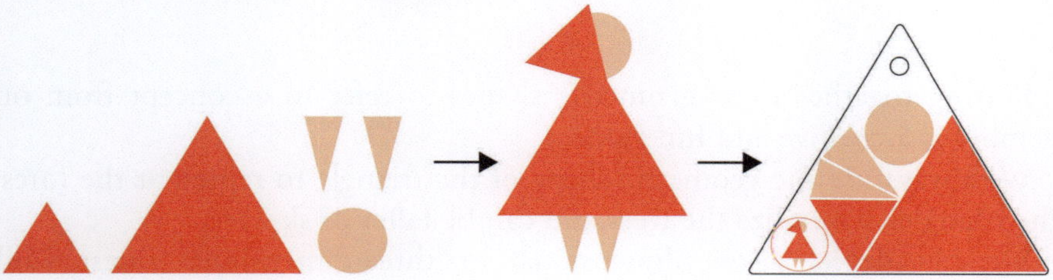

We have thus created a small puzzle of five pieces that can be inserted into a two-dimensional triangle kit, with which we can represent Little Red Riding Hood in her various poses.

This is Little Red Riding Hood walking.

This is Little Red Riding Hood bent over picking flowers.

Putting together these geometric shapes to refer to a concept from our memory is a creative and fun game.

We always use the geometric shape of the triangle to represent the forest. The triangles symbolize the trees and can be taller or shorter.

We can depict a two-dimensional or three-dimensional forest while preserving the fundamental elements of the selected basic geometric shape, leaving the rest up to our imagination.

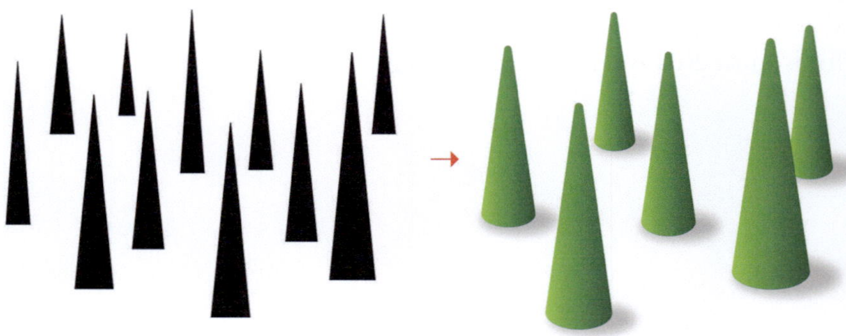

In a three-dimensional game, the trees can be made with cones. If we want to create a construction game, the cones can be stacked on top of each other, from the largest to the smallest, to grasp the concepts of size and sequence.

In a game with rules, the trees/cones can constitute the elements of a setting to be assembled, where Little Red Riding Hood, who is the player, must follow a path. We will also have a Blue, Yellow, Green, Orange, and Little Red Riding Hood… as many as there are players.

The following is how the fairy tale of *Little Red Riding Hood* can inspire educational games and themed toys.

If we are immersed in a dense forest and we do not see the tops of the trees. We can represent the forest also with tall rectangles, if the representation is two-dimensional, or with tall cylinders, if the representation is three-dimensional, which simulate the trunks.

Even this representation of the forest can stimulate the invention of a game. For example, a game of sequences, or of images to reconstruct by observing drawn cards.

If we consider the two-dimensional representation, we can create an educational game to learn the concept of front and back. If we add Little Red Riding Hood, we can place her in front, behind, or in the middle of the trees.

We can obtain the concept of far and near in the two-dimensional representation by inserting a smaller or larger triangle.

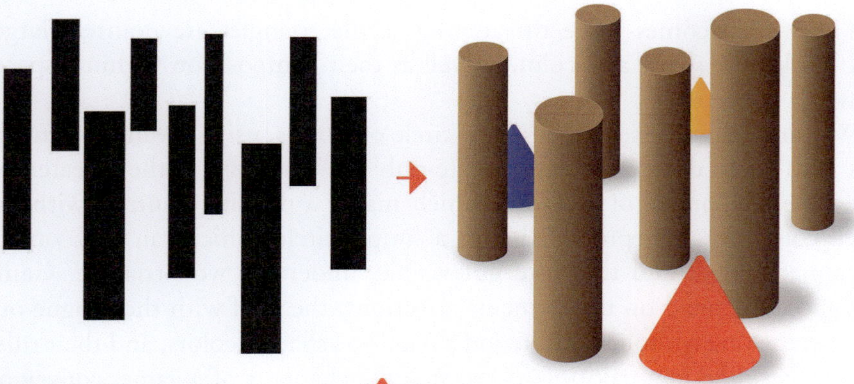

How do we now represent the wolf within this scene?

A simple illustration of this could be made with three triangles.

The wolf becomes more threatening as the triangles are pointed, large, and black. The shape and color, as well as their composition within a space, convey a concept.

We can imagine inserting it into a circle of wolves, using seven other triangles with seven different colors, where only the black one is the threatening one. The red tongue of the wolf, which makes a pleasing contrast with the black color, can be represented with a simple circle, which can stick out or remain included, and therefore not visible, under the wolf triangle. Game actions can depend on the elements' positions: the wolf with the tongue out can eat, the one without cannot, and so on. The shapes, colors, and their rules composites, like in a painting, determine the elements of a game, conveying concepts to which a set of rules can be added.

In the scene we are going to represent, the colors aid in avoiding confusion. If we choose black for the threatening wolf, we make the trees green, brown, or black, and Little Red Riding Hood will be red; other colors can be added if the game involves more players.

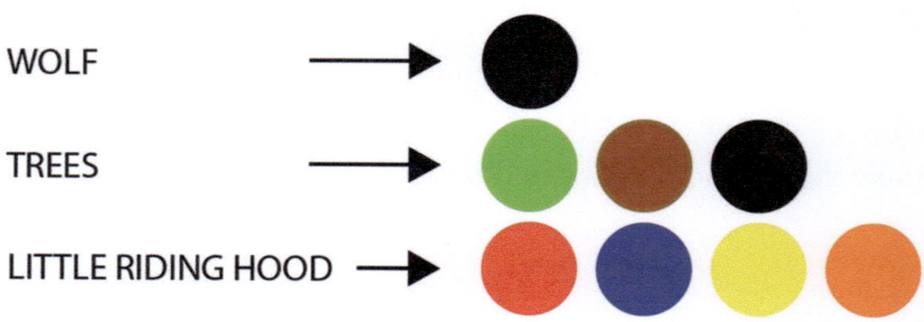

In this way, we'll be better able to recognize the parts that make up the scene. In an educational game, this aspect is important in order to clearly distinguish the components.

We can add additional details to the wolf's figure that may become significant elements within the game. If we add a red eye to the wolf, this detail immediately connects it with the red color of Little Red Riding Hood. If the eye is green, then there will be a green Little Riding Hood, and if the eye is yellow, also a yellow Little Riding Hood, etc. Certain elements can become the key to the game we are going to develop.

Other elements that characterize the wolf are the white teeth and a furry tail. How can we use these elements?

The white teeth stand out on the scene, acquiring great power if the color on which they are placed allows a strong contrast. In an educational game of image construction, backdrops of various colors can be provided to create emotionally different situations. The scene becomes more or less scary due to the colors, shapes, and sizes of the pieces and how they are positioned.

The images and settings that we are going to build in the games are an extension of the real world.

In the two-dimensional or three-dimensional representation of a scene, which can be precisely the scene of a fairy tale from which we want to extract our game concept, gravity has a very present physical and imaginative force. In terms of forces, we know gravity best, and it determines the shapes we want to represent. Horizontal shapes give us a sense of stability and calm, while tall vertical structures require energy and action to be knocked down. This concept can be the basis for the structure of a game.

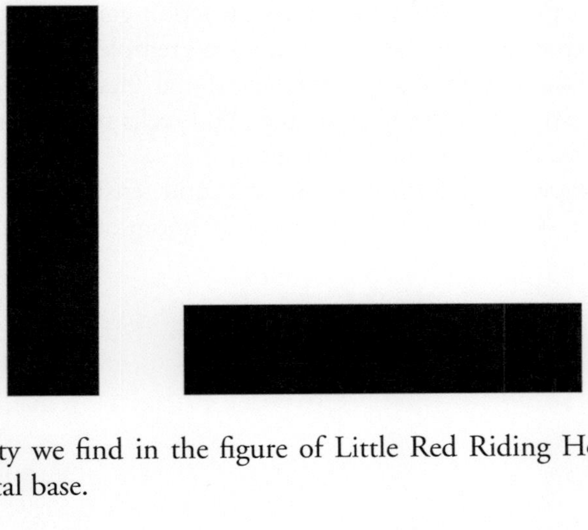

The stability we find in the figure of Little Red Riding Hood lies in its wide horizontal base.

If we put a horizontal bar between two verticals, we get stability.

This composition, created within a construction game, provides us with a sense of security similar to that found in Greek temples.

We get satisfaction in placing the horizontal element because it gives stability to the structure; in the game the child seeks stability and functional solutions, obviously corresponding to age.

Diagonal shapes instead imply movement and tension: they could be the beams of a roof, the branches of a tree, or the boom of a crane.

A diagonal is clearly visible in a mountain, in the roof of a house, and in a wave.

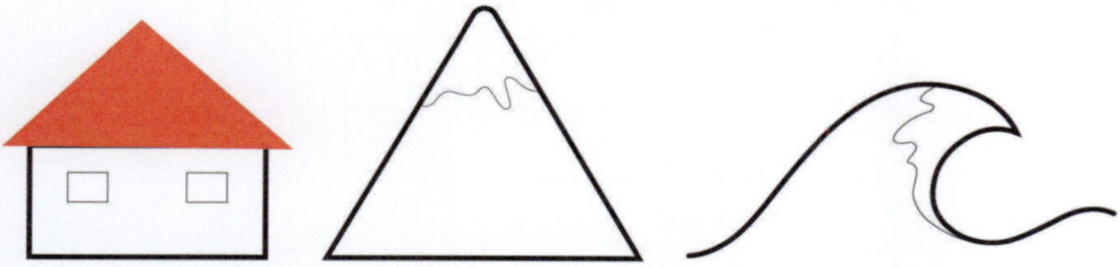

We can find many diagonals in the fairy tale of *Pinocchio* if we think about his hat, his nose, the fairy's hat, or the beard of Fire Eater.

These diagonals invite movement. Toys that involve movement and tension have recognizable diagonals.

Exercise 15

Choose a fairy tale and identify some of its characteristic elements or the main figures of the story. Associate with each element the geometric figure that represents it, as well as the most suitable color. For example, in the fairy tale of *Little Red Riding Hood,* we can write "Little Red Riding Hood" in the column of elements, "triangle" in the column of geometric shapes, and "red" in the column of colors.

Title of the Fairy Tale:		
Elements that characterize the fairy tale	Corresponding geometric shape	Corresponding color

Some fairy tales contain many characterizing elements, from which it is also easy to extract geometric shapes and corresponding colors, while in others this work is a bit more difficult, and the elements are few. It is not important how many of these elements there are; the important thing is to be able to make a selection and choose only the most significant ones. A game or a toy can be made up of many or few elements. In its compositional project, it is useful to identify the point of greatest attraction, considering that the point of attraction of an object is not always in the center.

What happens if this point of attraction is not in the center? The representation becomes more dynamic.

The axels of a pull toy's wheels do not necessarily need to be fixed centrally!

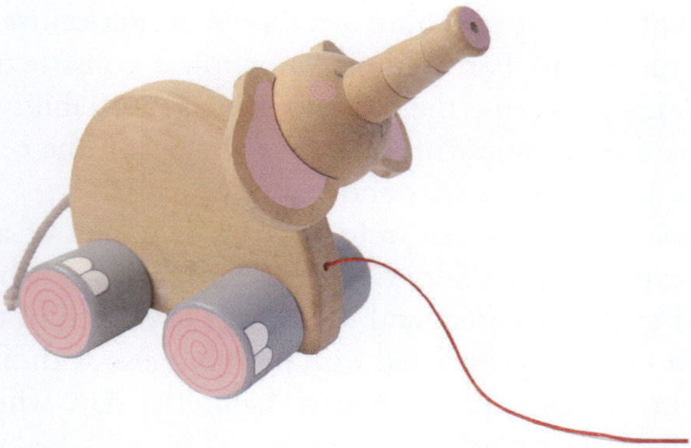

Pull-along elephant, *Zookids* line, Sevi/Trudi (2002)—design by Irene Guerrieri

If the attachment is off-center and asymmetrical, when we pull the toy, the animal tilts a little to the right and left alternately during its path. This movement, more similar to the natural movement of an animal, gives the toy greater dynamism.

Even when we draw the eyes of a puppet, we can obtain more static or more dynamic expressions depending on whether we position the pupil in a centered, off-center, or asymmetrical way.

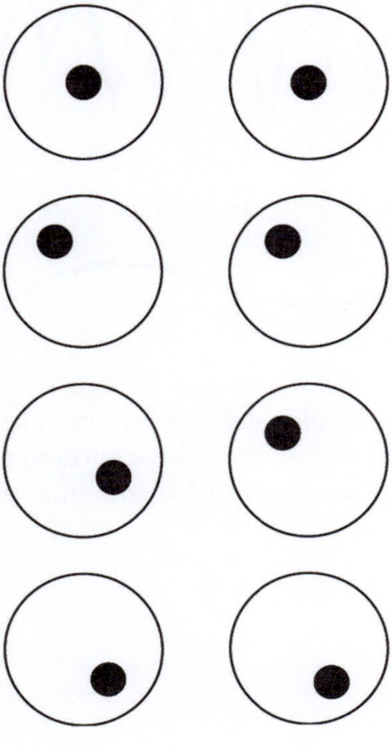

In the design of a toy, we can think of a central or non-central composition, depending on the needs. For example, in a tower construction game, the center is fundamental because the central setting gives stability and security.

As in the board of a game with rules, the image and the central structure of the path give a sense of security and stability.

Even the choice of colors can influence the dynamics of a game, so the colors that we extract from a fairy tale are crucial, and to each we can give a meaning linked to the dynamics and the correct operation of the game.

Red is a color normally associated with love, but also with blood, strength, and fire. A light background symbolizes light, the day, while a dark one represents darkness, the night.

The effect of color is profound. Yellow normally represents the sun, and blue, the sea or the sky. A red rose expresses more strength and love than a pink rose. A white swan appears purer than a speckled duck.

Returning to shapes, we can say that pointed ones are more threatening, while rounded ones are more reassuring.

We feel more frightened if we encounter pointed shapes.

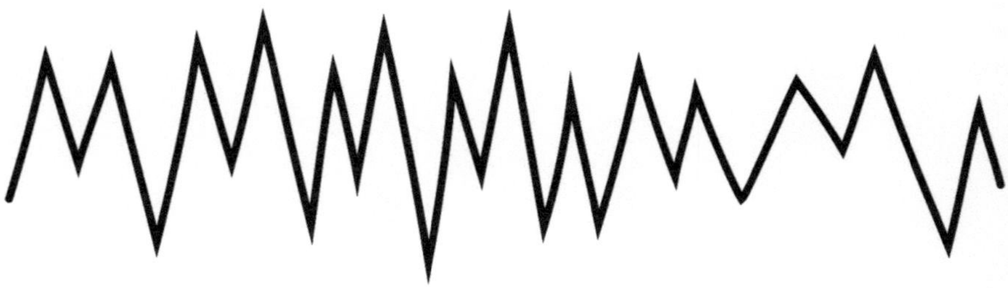

We feel safer and calmer if we see rounded shapes.

Considering all these factors, we can say that shape and color are at the basis of every conceptual choice and that the reference to a fairy tale helps us identify the purpose, the dynamics, and the operation of our project.

If the goal is to separate all the red shapes in a set into two groups, we could separate by shape.

If instead we have the same group of shapes but with different colors, we could divide by color.

Exercise 16

Using a pair of scissors and colored sheets, cut out the shapes representative of the fairy tale you have chosen.

Cutting colored paper into geometric shapes suitable for narrating a fairy tale can be a good starting point for devising a game. You can collect and develop ideas offered by the narrative. Assigning colors and shapes to objects and figures contributes to the narrative.

The memory and visualization of a fairy tale interact with other ideas, and new products are born. This is a creative and enjoyable process where memory and imagination collaborate harmoniously.

While deciding what to cut and how to do it, you must ask yourself many questions about how to better capture one concept over another.

"The fairy tale is the place of all hypotheses: it can give the keys to enter reality through new roads, it can help the child to know the world" (Gianni Rodari).

9

The Choice of Language

Through Games, Concepts Are Taught and Learned

Abstract The analysis of the form stimulates the birth of new concepts and defines the choice of language. Each figure, character, or object chosen as the inspiring motif of our project, analyzed in its form, can inspire new concepts.

Every figure, character, and object, chosen as the inspiring motif of our project, analyzed in its form, can inspire new shapes and new concepts, and help to create new figures or characters. It is a creative step that sparks new ideas and pictures, and is a key component of reading.

Analysis of form stimulates the birth of new concepts and defines language choices.

A significant example is given by the interactive book *La Rana Bastiana*, published by La Coccinella in 2003, which we have already mentioned in Chap. 6, concerning transformations.

I. Guerrieri, *Toy Design*, https://doi.org/10.1007/978-3-031-98872-1_9

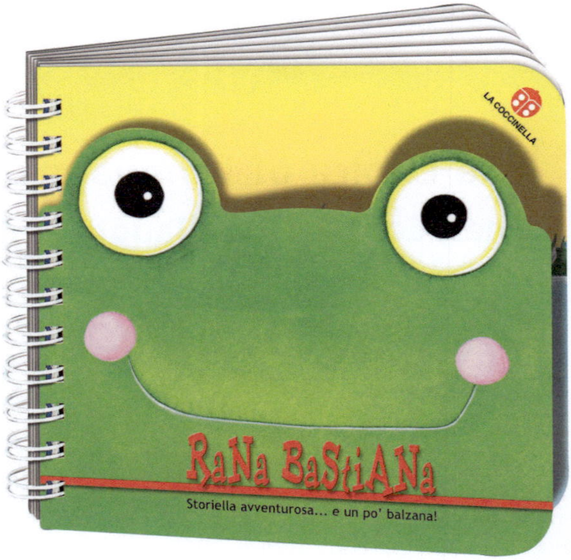

***La Rana Bastiana*, Libri Buffi series, La Coccinella (2003)—design by Irene Guerrieri**

Plitsch, Platsch is a game born on a rainy day while I was watching the raindrops falling on the ground outside. I analyzed the circular shape of the falling drop and considered representing it with three concentric elements that vary in size and shades of blue. In this formal analysis lies the substantial choice of the adopted language and the *concept* that gave life to the game. The German company Magnetspiele, which specialized in games with magnets, wanted to introduce the magnet to my idea, and with them was born *Plitsch, Platsch*, which went into production in 2008. The raindrop is associated with the rainbow and the different weather conditions; from these different associations, the game dynamic is born.

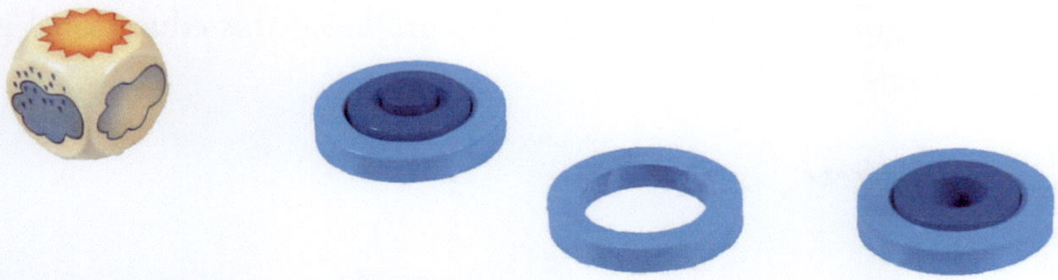

Plitsch, Platsch, **Oberschwabische Magnetspiele (2008)—design by Irene Guerrieri**

In *Opposites*, the opposites create a relationship and a match, giving birth to the game language. *Opposites* is a interactive book with wooden pages and removable objects. Form and content define the mode of play. On the idea of "opposite," I built the concept.

Opposites, **Innovation in a Box (2013)—design by Adittee Naargolkar, Irene Guerrieri**

In *The Bigger & the Smaller*, each animal is associated with its color and its size, according to a scale that goes from the smallest to the largest. Here, the choice of language is attributed to the color and scale of size. I extracted a concept by putting several animals in relation to each other.

The Thai company Innovation in a Box produced this educational game for kindergartens in 2013.

***The bigger & the smaller*, Innovation in a Box (2013)—design by Irene Guerrieri**

Often the choice of language is made through the main attribute of a character, which can materialize in an object, such as in *Popeye,* where the concept of strength is associated with the can of spinach. Many objects endowed with their own attribute remain in our memory in their inseparable duality, like the rabbit with the carrot or the dog with the bone.

Exercise 17

List other objects with their corresponding attributes.

Object	Attribute

Associating an object with its attribute means expressing a concept, making a choice of language with which we want to communicate.

Through attributes we can invent many things and give life to real communicative languages, which, through play, find their application.

Another effective way to do this is to extract an object of daily use from the banal and invigorate it with a new meaning. For example, the notion that a knife can be used to identify an aircraft.

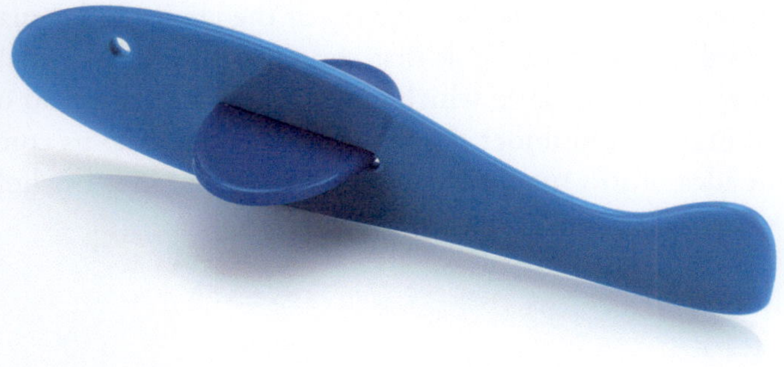

Knife, *Traffic line*, Fratelli Guzzini (2017)—design by Irene Guerrieri

You can make up stories for children by giving new meanings to the objects on the table or highchair. The child, if free to act, spontaneously invents fantastic binomials: thus, flying dishes, meatballs that sail in the soup, and peas that sink into the glass are born. Choosing to design a knife that turns into an airplane means choosing to speak the same language as children.

Attributes, associations, and new meanings are all tools that help us create the language to make the game become the means to learn concepts.

Collodi extracts his Pinocchio from a log of wood; then, when he enters the land of toys, Pinocchio turns into a donkey, and at the end of the story, he becomes a flesh-and-blood child.

The fairy tale is often used to attribute new meanings to convey concepts. Always considering the fairy tale of *Pinocchio,* the nose that lengthens and shortens is itself a game contained within the fairy tale, which works following certain rules. Pinocchio's nose is a fun object, made of wood, which can be very well imagined, to which a new meaning has been attributed, but that serves to convey a concept. Through Pinocchio's nose, a language was created.

In Andersen's fairy tales, we notice a particular mastery in animating the most banal objects by creating new languages. Just think of *The Princess and the Pea*, *The Little Mermaid*, or *The Tin Soldier*, which contain within them real games.

This means that we can draw many project ideas from classic fairy tales.

Attributing new meanings to objects, people, and animals allows them to create the language, but they are not always strictly logical interventions. As a result, an invention often arises with a new logic. To get to this, you jot down, sketch, or doodle a bit before arriving at the choice of language.

Take the Moon is a bedtime book that contains an object: a small slice of moon that lights up in the dark, which can be hung with a thread or carried as a lucky charm or placed under the pillow at night to keep us company.

Around the idea of this small slice of moon to thread into the holes of the cardboard pages, the book was developed.

***Take the Moon*, Gribaudo (2018)—design and illustrations by Irene Guerrieri**

***What I don't like*, Gribaudo (2017)—design and illustrations by Irene Guerrieri**

Gribaudo's two interactive books, *What I Like* and *What I Don't Like*, are about Nina and her daily adventures. These books exhibit the effectiveness of holes. Games and language therein are based on the use of these holes. You stick your fingers in them, and the story comes to life. The little reader becomes a protagonist alongside Nina. The language is defined by the movement of the fingers inside the holes, which refers to daily actions that Nina loves or does not love to do.

10

Child and Adult

The Toy as an Imitation of the Adult World and a Tool to Retain the Adult

Toys play a key role in the development of personality, especially emotionally, activating mechanisms of identification and projection.

There are toys that imitate on a smaller scale the adult world and those that require the adult's attention. In both instances, the toy symbolizes the world against which children evaluate themselves while also serving as a projection and extension of their identity. Children are naturally receptive and want to imitate and learn from adults. Their need to imitate the adult is part of their will to grow.

The toy industry miniaturizes the adult world, manufacturing trains, cars, dolls, and doll clothes, as well as games in boxes such as *The Little Explorer* and *The Little Chemist*. Everything that belongs to the achievements of the adult world becomes a toy for the world of children.

Those who play with dolls and their clothes, furniture, pots, and miniature houses project into the game all their knowledge about domestic life. They cuddle the doll to express their need for affection and scold it with the same words with which they are scolded. Toys enable children to manage the elements that represent their desires and fears. Through the toy, a mechanism of identification and projection is activated so that children express feelings and gestures of love or aggression. They measure themselves with the object so much that they sometimes want to disassemble it to see how it is made or even destroy it.

Often, while playing, a child will engage in self-dialogue, explaining the game and bringing things to life.

Toys establish connections between adult society and child society, highlighting obvious affinities. In this relationship between the adult world and

I. Guerrieri, *Toy Design*, https://doi.org/10.1007/978-3-031-98872-1_10

the child world, it is also important to highlight the fact that often, children seek the adult in the game; they need them, because the adult has a wider experience, and the game is enriched and opens to new horizons.

Even in cultures hugely different from ours, or in antiquity, playful activity within a community has always highlighted the especially important relationship between the adult world and education.

In history, there is then an important distinction to be made between male and female education, where the toy offers the child a distinct instrument for emulating the adult environment, regarding gender.

In every age and culture, children have always produced their own toys by imitating and reworking things seen or seen done by other children or adults. In poor societies, in fact, it is still a widespread custom, especially in those countries far from the spread and consumption of industrial toys.

We can indeed talk about toys both as a handmade product and as an industrial product: in the former, there is often a greater involvement of the adult.

Two phases characterize the development of the toy in modern society. The first phase, which goes from the beginning of the eighteenth century to the first half of the nineteenth century, sees in the development of the handmade toy the application of the first ingenious mechanisms; the second phase, between 1820 and the end of the nineteenth century, is characterized by a very refined aesthetic production that defines a real poetics of the toy: think of dolls with their houses rich in details or the steam engine applied to some miniature vehicles.

We can then add a third period, between 1880 and World War I, which sees the first great affirmation of the toy, especially in Germany. Toys primarily use materials like tin and wood to depict the many aspects of the adult world.

From this period is the invention of the *Meccano* (1906), by the Englishman Frank Hornby, which is the first modern construction toy characterized by metal pieces that can be assembled together with screws and bolts.

Between the two wars, Germany confirms its supremacy in the production of toys both qualitatively and quantitatively.

Meanwhile, in America, the American Charles Darrow invents the *Monopoly*, a real, playful representation of the functioning of capitalism.

After the Second World War, plastic increasingly replaced wood and metal as basic materials, while batteries, electrical circuits, and remote controls took the place of traditional spring loads. Amid economic advancement, novel elements are integrated into toys. Missiles and robots emerge, alongside distinctly more streamlined and inventive dolls that embody contemporary lifestyles.

Between the fifties and sixties, two very innovative toys appear destined for worldwide success: these are *LEGO* and *Barbie*.

The *LEGO* brand synthesizes the two Danish words *leg* and *godt*, which together mean "play well." Conceived by the Englishman Hilary Page, it was perfected by the Danish carpenter Ole Kirk Christiansen. The prototype of the plastic toys were made at the artisan level in wood. As it was produced in solid plastic and bright colors, the toy gradually replaces Meccano, which has a cold and metallic structure.

LEGO shows extraordinarily creative stimuli on the educational level: with the addition of a series of special pieces, you can connect to a computer using the program *Logo,* created in the sixties by Papert. *LEGO Logo* allows the child to design toys, develop and verify hypotheses, solve problems, and check the results. Learning becomes an active construction process based on concrete experiences.

Barbie was introduced as the inaugural model in 1959, created by Ruth Handler. The concept originates from cardboard dolls that girls would cut out along the figures' contours, which come with several costumes that can be put on by folding the corresponding tabs.

With *Barbie*, the concept of the doll radically changes. *Barbie* represents the model of the American woman in the fullness of her physical form, between twenty and thirty years old, who can satisfy all her desires, represented by a myriad of clothes, accessories, and furnishings. It represents the progressive emancipation of the female figure in the welfare society. *Barbie,* with millions of copies sold worldwide, has made its manufacturing company, Mattel, a world leader in the toy industry.

The industrial production of dolls begins in Italy at the end of the nineteenth century with the Furga Company, but the take-off of the toy industry in the Apennine Peninsula occurs after World War I. The production of Lenci dolls marks this transition. The innovative idea was to use padded and pressed cloth for the production of dolls, which soon became famous in Italy and abroad. In 1922, the brand *Lenci* was registered, an acronym for *ludus est nobis constanter industria.*

The concept of toys has evolved significantly since the advent of industrial mass production, incorporating advanced technologies and design methodologies, innovative materials, the establishment of safety-oriented regulations (e.g., EN71), and the enforcement of copyright protections, among other developments.

Today, toys are subject to rapid evolution over time, and they are influenced by temporary trends: some types of toys have been on the market for a few years, others only for a few months. Toys tend to reflect the culture and customs of specific historical periods. However, some categories of toys always seem to be in fashion and probably are part of the cultural heritage common to various populations. This applies both to toys whose origin is lost in history, like dolls and soldiers, and to others of more recent origin, such as toy cars. Even in this case, it is possible to identify an evolution, linked, for example, to the choice of materials.

The *yo-yo* is a game that dates to antiquity, but even today we find it in very sophisticated versions. Between the seventies and eighties, the *yo-yo*, supported by technological development, became luminous.

Recently, toys have also been marketed as gadgets associated with movies or merchandise brands. What is important to highlight is their role in constantly connecting the two worlds: the adult and the child.

In *My Fashion Doll*, the kit includes the packaging of the little dress with all the necessary elements to make it, except for the needle, which for safety reasons cannot be included in the box. The activity requires the support of an adult, who can accompany the child in achieving his goal, supervising, and helping him in using the needle to sew.

My fashion doll **(2013)—design by Irene Guerrieri**

***Grandmother & Grandfather*, Beleduc (2012)—design by Irene Guerrieri**

Even in the fantastic world of puppets and marionettes, which sees the child as both a spectator and a protagonist, we can highlight a close relationship between the adult world and the children. In the first case, the presence of the adult who animates the puppet show is essential; in the second, the child becomes the protagonist but uses the script to converse with the adults and communicate to them, through the character the child is manipulating, things that he would never have had the courage to say alone. Schools use puppet-marionettes to teach, tell stories, speak English, and even act as teachers. To the puppet-teacher, children often say things that they would never have dared to say to their real teacher.

Some children, while manipulating the wolf puppet, distance themselves to avoid being eaten. They are inescapably the ones involved with the wolf's fabric, yet… uncertainty persists.

***The Puppet Theater*, Pirondini Italia (2002)—design by Irene Guerrieri**

11

The Illustrated Toy

Illustration as an Integral Part of the Game Object

Abstract If in a book the illustrations suggest and accompany the narration, in a theater they become an integral part with the elements that compose it. The illustration can often become an integral part of the playful proposal, whether this is a small theater or an illustrated box that itself becomes an integral part of the game.

From the puppet theater designed for Pirondini, the idea was born to create a certain number of shapes in natural wood on which to then apply the illustration of the character you want to represent. This choice has simplified production, adding an activity to the kit, with the educational goal of involving children in the painting completion of the figures. We cannot imagine the shapes of a theater without the illustration that refers to a character. The theater can only work if it is illustrated and set in the story you want to represent. In addition to the neutral kit, the company found it useful to offer separate kits that are already illustrated based on the themes of the most famous fairy tales. The company thus invited me to illustrate the figures from the fairy tales of *Little Red Riding Hood*, *Cinderella*, and *Snow White*.

If illustrations in a book suggest and accompany the narration, then in a theater, they become an integral part of the overall performance elements.

If, until now, we have focused mainly on the analysis of form and color as essential communicative elements, now we introduce the illustration as an integral part of support and completion of the game proposal.

I. Guerrieri, *Toy Design*, https://doi.org/10.1007/978-3-031-98872-1_11

Empty silhouettes of puppets, Pirondini Italy (2002)—design by Irene Guerrieri

The contribution of the illustrator often becomes indispensable in the design of a toy, although there are also aesthetically minimalist toys on the market, where shape and color exhaust the communicative part.

Illustration, when required, serves as a practical and essential component of the toy, encompassing both the object and its packaging.

It is essential to acknowledge the contribution of Attilio Mussino (1878–1954) to the Italian toy industry in the 1920s, as he was one of the foremost illustrators for children. The Cardini firm in Omegna conceived a remarkably innovative packaging concept by commissioning Mussino to design the toy boxes, which, when opened, became integral components of the toys, converting into a house, hangar, port, airplane, ship, or train. These products, particularly their packaging, were promoted in the pages of the *Corriere dei Piccoli*.

Cardini toy advertisements—private collection of Giuseppe Zappalà

Illustrated Cardini toy—private collection of Giuseppe Zappalà

The illustration plays a significant role in the narration. Additionally, we can say in a more engaging and concise manner that the illustration transforms into a game.

In recent years, there has been a significant development in the market for board games. Stores are filled with games that are no longer restricted to the likes of *Monopoly* and *Risk*. In books, illustrations convey ideas, and in board games, they depict settings that are impossible to misunderstand.

As far as commercially available games go, Dixit best exemplifies the role of the illustration in development dynamics. It was released in France in 2010, which, thanks to its success, has already given life to seven expansions. Many game publishers rejected this project because the author used images taken from children's books, which created a copyright issue. Roubira, the author, then looked for an illustrator.

He was introduced to Marie Cardouat, a children's book illustrator, and in the end the game was produced by Libellud, a publishing house that was born thanks to *Dixit*.

During the release of the subsequent expansions, the publishing house, demonstrating good intuition, changed the style of the illustrations to make the game more varied.

Here we understand very well the importance of illustration in the toy. In this chapter, we present the illustration as a crucial component of a fun proposal's narration and completion, shifting our focus from the analysis of shape and color as vital communicative factors to their role altogether.

Not only do children's book illustrators design the characters that go along with the narrative, but their work is also crucial in the creation of toys.

In many of the toys I have designed, the illustration was born alongside the shape, and in certain cases we can say that it adapts to the shape; or vice versa: the shape is influenced by it.

In the wooden puzzles designed for Sevi/Trudi, this fact is especially evident.

Book, puzzle, and inserts, line *Zookids*, Sevi/Trudi (2002)—design by Irene Guerrieri

12

Safety

Toys Must Be Safe Objects

Abstract The increase in the production of toys has led to the problem of safety. The law establishes that toys can only be placed on the market after receiving a recognizable certificate. How to achieve it?

The significant rise in toy production and consumption has resulted in safety issues.

Children, with their exploratory behavior, seem to have no sense of danger: they learn through trial and error. They are attracted by flavors, smells, and contrasts between colors; they put objects in their mouths (0–3 years), and they insert their own fingers and hands everywhere to explore.

The law states that toys can only be placed on the market following receipt of a certificate, recognizable by a CE mark, with which the manufacturer or their representative certifies under their own responsibility that the toy has been manufactured in compliance with safety standards.

The CE mark means European Conformity, and in concrete terms it is the graphic demonstration that the product to which it is applied complies with the regulations in force within the European Community that govern the production and market introduction of that product. The CE mark, if applied legitimately and not erroneously or fraudulently, guarantees that in its design and manufacture, the product has been in total compliance with current regulations.

There are specific bodies, called safety institutes, that issue the CE mark on toys after subjecting them to a series of tests that verify their safety level

I. Guerrieri, *Toy Design*, https://doi.org/10.1007/978-3-031-98872-1_12

and that concern physical and mechanical properties, flammability, chemical properties, electrical properties, hygiene, and radioactivity.

Safety institutes provide suggestions on different prevention strategies to introduce safe products to the market. Risk reduction strategies are classified into two categories: active and passive. The first category necessitates consumer activity, such as heeding warnings; the second pertains to product design that eliminates the need for extra warnings, so ensuring protection without requiring specific actions from the user. The secondary strategy predominantly engages the designer.

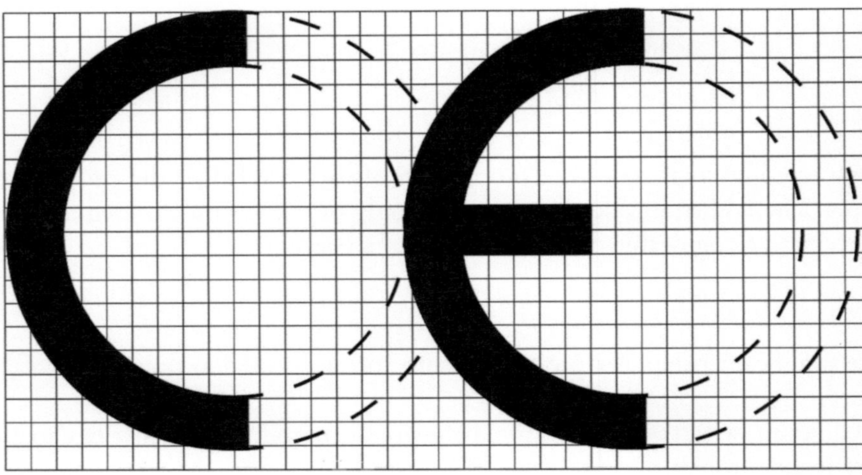

CE mark

It is important to underline that those who import into the European Union products coming from third-hand countries are subject to the same obligations as European manufacturers and therefore must affix the CE mark on the products. It is always necessary to check that the marking is affixed according to the graphic symbol provided by the regulations.

On many products imported from non-EEC countries, the marking is affixed solely to deceive the consumer.

Thus, the marking can be found on non-compliant products.

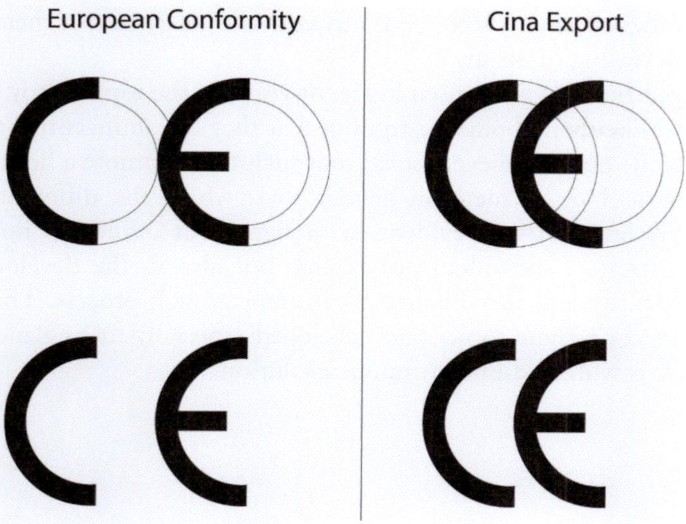

Example of non-compliant CE markings

Among the warnings for the consumer is an indication that a toy is prohibited for children under three years of age.

It therefore becomes essential to establish an age classification based on the characteristics of the toy. This will determine which requirements the toy must meet and which warnings it must carry.

The standards to be adhered to in order to introduce a toy that meets safety terms are in EN 71, which concerns physical–mechanical requirements—such as small parts, protruding parts, small balls and spherical toys, markings, migration of certain elements—as well as electric toys.

Leaving all the regulations to the competence of the institutes in charge, I want to conclude by making a small mention of some themes that recall the psychological and pedagogical aspects of toys.

The number of parents seeking to buy educational games for their children has been increasing markedly.

A strong debate has also arisen in recent years on the topic of toy weapons; some would like their abolition, arguing that they are an incentive to violent behavior, while others believe instead that such toys promote a healthy outlet for aggression. Not to mention gender toys, where the differentiation of sexes has always been very evident, with significant influences not only in the construction of individual personality, but also in the development of children's identity and social behaviors in their growth process. The cultural debate opened on these topics and developed especially in England and the USA, and it is wide and in continuous evolution.

13

The Realization of the Project

Dialogue with the Manufacturing Companies

Abstract Designing also means finding the solution to the problems that gradually arise. It is often necessary to involve other professionals and skills to find the right solutions. I thought of interviewing some key figures of some Italian companies well known also in the rest of the world to bring their point of view.

The realization of an idea presupposes the execution of its project. Communication with the manufacturing company is crucial in this step, involving several professionals to render the project practicable. The problems that arise from it, both technically and economically, are generally numerous.

Designing is essentially finding the solution to the problems that arise from time to time. As in life, problems can sometimes seem insurmountable, but if they are faced one at a time and involve other skills, solutions can be found.

The solutions can sometimes coincide with small changes to the initial idea, but other times it can give rise to a real new invention.

The execution of a project necessitates a sequence of essential actions organized in a logical order dictated by field experience.

The designer who proposes an idea and addresses the challenges of execution recognizes the necessity for the intervention and experience of additional professionals. Thus, a team of workers is created, who, at various stages of product development, discuss and make changes to the initial idea. Each skill contributes to the successful realization of a product, and the final result is the fruit of a comparison and exchange of experiences.

I. Guerrieri, *Toy Design*, https://doi.org/10.1007/978-3-031-98872-1_13

An ostensibly "brilliant" concept that seeks to manipulate method relentlessly to attain an exceedingly challenging, costly, and potentially impractical yet aesthetically pleasing outcome will invariably fail to prosper.

In this regard, I considered interviewing key figures from certain companies that are important to me and have already been mentioned in various parts of this book.

I asked Domenico Caputo, owner of the publishing house La Coccinella:

What are the main characteristics that an interactive book proposal must contain to be chosen?

Domenico Caputo—The characteristic that a project we can consider must contain is, *primarily*, the originality of the graphic-content structure, in line with the stylistic features of the publishing house. As a second aspect, the playability of the book on which and in which the child can interact is stimulated by mechanisms contained in it, and as a third point, the possibility of conveying an educational message in the form of a game helps the child in acquiring concepts or skills useful for his growth.

How many proposals do you receive from outside, and how many proposals are born within your team?

Domenico Caputo—The projects we bring to completion are largely developed within our structure, engaging our dedicated staff. Those that come to us and that we choose from outside are almost entirely of Italian origin. Throughout the years of the company's existence, we have completed only a small number of foreign projects.

What are the figures involved in the choice of a product?

Domenico Caputo – Predominantly a creative with good experience in developing products for early childhood and with good visual communication skills. It would be better if he combined creative skills with those of an illustrator, with graphic-artistic characteristics in line with the stylistic features of the publishing house. Normally the text is developed internally, and, where the project requires it, we rely on an author of proven professionalism and experience in the world of childhood. Sometimes, and always only on specific projects, we entrust ourselves to pedagogues with the above experience, consultations, or validations of more specifically scientific themes or topics.

I then turn to Federica Monzio Compagnoni, *toy designer* of the German company Beleduc, producer of educational games for children:

How does your company test a product before it is definitively chosen for production?

Federica Monzio Compagnoni—From the *concept* phase, all our games are born and developed according to the motto "children do not play to learn but learn because they play": our products are therefore characterized by an intrinsic educational value; each promotes learning in one or more educational areas. The care of this aspect does not only derive from the work of the designer, but from his close collaboration with other professionals. The different types of products are tested in kindergartens in various countries around the world; we want to ensure that our toys are suitable for everyone, without language or cultural barriers. Our designers personally go to nurseries and kindergartens, proposing game prototypes to children and observing their reactions and ways of interacting with them. This direct observation is a tool that provides important feedback on the toys under analysis, suggestions for improvement, and new ideas. The observation is carried out simultaneously with the teachers, who at the end of the playtest session provide detailed and extremely useful evaluations, a fundamental and professional contribution with a focus on the more concrete educational aspect in relation to the abilities of children in the different stages of development.

For some products and collections, we also collaborate with university professors, pedagogues, experts in the field of education and professionals in various specific fields who provide us with comments and advice, opinions, and specialist opinions on our games in order to perfect the educational aspect.

Parallel to these activities, we also turn to families: from tests within family units, we can obtain other very useful contributions, which help us to further improve the product.

Only at the end of these tests and after having collected various feedback do we finalize the product based on the results obtained; we then proceed with the subsequent stages of development, up to production, certain that the game is suitable and approved by various experts, not least the children themselves!

I asked Valentina Gottardi, owner of Cocai Design, a company that produces tools for education about nature and beauty inspired by the Montessori Method:

How is a Cocai Design product born, and how much time does it typically take from the birth of the idea to its introduction to the market?

Valentina Gottardi—The inspiration for creating a new product can come from various sources: from nature, from children, from our childhood, from a walk in the woods, from the band of chickadees that perch on our trees in the garden…

We are passionate about the natural world, especially the details we often overlook. The principles that inspire us are those related to the Montessori Method and the examples of Bruno Munari: freedom, autonomy, creative thinking, and a few well-defined constraints. We believe that in educational games there should not be a "right way" or a "wrong way" to play; the important thing is the learning process and the stimuli that derive from it.

Cocai's products are diverse. A "simpler" product, like a nomenclature, needs about two months of work. This period includes the choice of subjects (in collaboration with our naturalists), the creation of illustrations and packaging, corrections, printing, and packaging.

To produce a more complex object, like a carved leporello, we worked for more than a year, from the conception to the creation of drawings and files, from the search for suppliers/producers to fundraising, until the realization.

I inquired of Davide Francioni, project manager of Headu, a company that manufactures games, toys, and instructional books for children with a unique pedagogical approach:

What criteria does your company use to choose a proposal to carry forward, and who are the figures involved in this process?

Davide Francioni—A proposal must be consistent with the company's editorial plans, both in terms of quality and economic sustainability. We carry forward the most innovative ideas in the educational and didactic field and those that have a good *design-to-cost* ratio. In this process, the *product managers* of Research & Development are involved, who elaborate the proposals of *game designers* and education professionals or develop projects also based on their own ideas, although the final decision is always up to the marketing direction, which effectively coordinates the entire process of defining the assortments.

What are the most common problems you encounter in carrying forward the development of a product?

Davide Francioni—Generally, a good idea might not be easily achievable for many reasons: the problems can concern editorial and marketing consistency, or they can be related to the economics of the project. However, the activity of design and planning has the function of minimizing the possibility of unexpected problems arising. At least, that's the goal!

For Luca Sassi, owner of the publishing house Sassi Junior, was asked:

How did the idea of combining the book with the game come about?

Luca Sassi—The idea comes from wanting to offer a dual opportunity to children: to learn and play at the same time, making the learning process something spontaneous, playful, and fun.

What are your most successful products, books, and games on the market, and why are they successful?

Luca Sassi—The most successful product is perhaps our "Travel, Learn, Explore THE EARTH," a suitcase containing a 205-piece puzzle and a booklet that explains to children all the main characteristics of our planet, both from a physical-geographical and cultural-anthropological point of view. Though it is a simple method for kids aged 6–7, it has been valued for its educational and entertainment value. We have translated this title into many languages and printed more than 200,000 copies in 6 years.

Architect Giovanni Fumagalli, *designer* and *art director* of GAM Gonzagarredi Montessori, was asked:

What importance do playful aspects play in the design of furniture for children, and what does the *concept* of playful furniture consist of?

Giovanni Fumagalli—Giving a playful character to a piece of furniture, I don't think it meets a need of the children.

Sure, at least for a while, children love furniture designed for them, those at their size, which identify their space. Then that preference gives way to interest in other furnishings, often not made for them, but that they can experiment and explore, perhaps secretly, like the bed and the grandparents' wardrobes. Children know how to do things on their own.

Bringing a piece of furniture into the realm of play is a desire among adults.

It is part of their way of taking care of the little ones, of relating to them. In this sense, it is a positive, serious, and anything but easy commitment.

Bringing furniture closer to games involves a reflection on the idea of play and its multiple aspects. We must overcome the most common banalities, such as inflating colors and decorations. A deeper mechanism must be sought, which has to do with pleasure, evocation, and creativity.

For me, designing a piece of furniture with a playful character means, above all, breaking its functional univocity, giving it the ability to surprise, and opening it to alternative interpretations. It means starting a game with the user, a game in which I can only make the first move: trying to spark an interest, to intrigue, to provoke even different uses than the usual ones for that type of furniture. I can design a chair that can also be used lying down or upside down, one that combined with others forms a kind of bookcase, a table that suggests becoming a hiding place… Then, if it works, it's the children who surprise me by handling that piece of furniture in ways more creative than mine, totally unexpected.

I finally turned to Gianni Romandini, marketing manager of Fratelli Guzzini:

What is the process that your company must go through before approving a design proposal that comes from outside, and who are the people involved?

Gianni Romandini—In Guzzini, design adds value to the product and the designer, working with the intent to reinvent the object, participates with the company in the success or failure of the project. The product innovation process is transversal to the entire organization, involving the main departments: Product Marketing, Research and Development, Engineering, and Quality Assurance. Once the designer is commissioned, aesthetic proposals are received and selected; subsequently, the technical and economic feasibility of the project is verified before the resolution that starts the realization of the executive drawings and production equipment.

14

Asian Production

Where Most of Our Toys Come from

Abstract In this chapter I report my experience of my travels in Asia in contact with suppliers and the synergy that was born with a Thai company producing educational cloth toys and my friendship with the Indian designer Adittee Naargolkar.

In 2005, I journeyed to China, commissioned by Thun to develop a new children's line that the firm, renowned for its ceramic manufacture, sought to incorporate into its catalog, integrating materials such as wood and cloth with ceramics. The company asked me to create a line of toys that included about forty-five new items, divided into two categories: a soft series and a wooden series. The chosen animal subjects (elephant, penguin, fish, dog, and rabbit), identifiers of their corporate image and already present in their ceramic products, had to be proposed again as toys and accessories for children, to be made in soft padded fabric or in natural wood decorated in some parts. The chosen suppliers were Asian, for cost convenience and for the high production numbers expected. The experience for me was unique and unforgettable.

I had the opportunity to meet a culture and a production organization very different from ours, learning a lot from the people with whom I worked during that experience and with whom I then continued to stay in touch. I learned how important it was for them to receive drawings and ideas from me as a Westerner, because in the West the Asians look for the customer, and from the West, they want to learn mood and style. They viewed me, a Western designer, as someone from whom they could draw extensively and

I. Guerrieri, *Toy Design*, https://doi.org/10.1007/978-3-031-98872-1_14

grasp diverse concepts; I saw in them a truly unique willingness and perspective. A prototype was modified many times, and this process required long waits on my part before I would be able to see the results and provide any new directions. I was not allowed to work closely with them, as had been the case in Italy at the Sevi carpentry in Val Gardena. In China, you must wait, drink tea, take a tour of the city center, and only when the prototype has been completed are you allowed to show up at the factory, analyze the work, and provide details for any amendments required for the subsequent draft.

Thus, an intense week of tests and corrections is spent. "Hard work," the Chinese would say, but it yields remarkable results due to countless acts of immense patience and meticulous precision. The prototypes of the cardboard boxes housing a product were crafted manually, exhibiting precision and care. A meticulous task, devoid of haste, characterized by countless hours of dedication and unwavering attention, along with respect for the Western clientele.

I remember the scent of tea infused with flowers that bloomed in the hot water, which is also a symbol of the infinite patience of a people who are extremely dedicated to their work.

After my experience in China, I moved on to Thailand and then India: contact with Eastern culture has distinguished my profession.

In 2013, in Thailand, I began an important collaboration with the company Innovation in a Box, whose owner was Adittee Naargolkar; together, we created many educational games intended for Western kindergartens. With Adittee in 2014, we created a brand, GUNA, from the first two letters of our surnames, Guerrieri and Naargolkar. This was a fascinating experience that involved me in the design and illustration of innovative concepts. In Thailand, the company was all-female: fifteen artisan women, diligent and attentive to their work, who used machinery that was obsolete to me but functional, and with which they produced very high numbers of toys.

Adittee is an Indian woman, although her company was in Thailand. Her energy, courage, and determination have consistently impressed me and imparted valuable lessons. Rich in resources and inventiveness, Adittee represents for me the Eastern woman rooted in her traditions, yet simultaneously eager to emancipate herself and to approach in some respects the Western world. She found in me the person with whom to unite to give visualization to her ideas, and I found in her the figure, with pedagogical and productive experience as well as great relational skills, that I was looking for.

Adittee Naargolkar (2013)

In 2014, we participated in the Nuremberg toy fair with our own stand, thanks to the synergy that produced a rather wide range of articles.

My collaboration with the Asian production world continued to develop in numerous other contacts with wood and fabric manufacturing companies, with which I successfully established a modest range of products under my brand. These were all enriching and unique experiences, where "the day never ended," especially when working for an American client; and while you were falling asleep, someone else was waking up on the other side of the world, the work never stopped; it was taken up at accordingly by people in different places who remained in touch throughout. Each person, with their own culture and training, was called to add to the new product from their own experience and heritage.

I have felt first-hand that the world is round and that life continues in the constant alternation of day and night, without interruption; when it is night on one side of the world, it is day on the opposite, and a good idea never goes out if you really believe in it.

Conclusions

Every morning when I get up, I look out my window at this view. It is the area of mountains that surround Lake Iseo, which I can admire from my house.

I recognize a giant face in the profile of those mountains. As if under that line of separation between the earth and the sky, a large figure with human features is hiding, resting, while I am about to start my day. "I wonder what he did during the night if I get up and he sleeps now?" I think to myself, and then I turn to the giant who rests silently.

I. Guerrieri, *Toy Design*, https://doi.org/10.1007/978-3-031-98872-1

THE MOUNTAIN GIANT

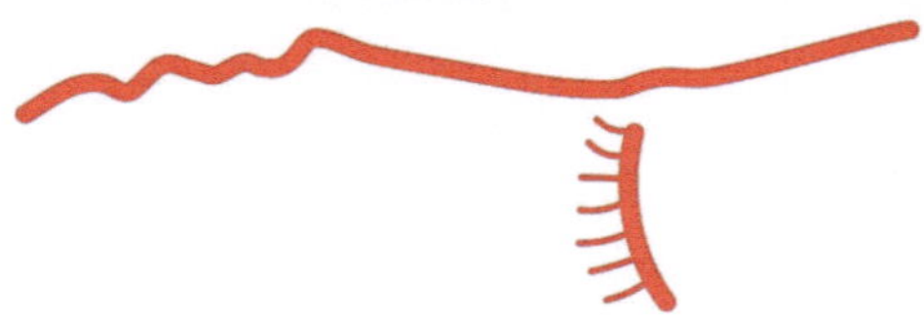

This is the game I repeat every morning. I never tire of repeating it and fantasizing about it, adding details to the spectacle that presents itself to my eyes every morning, always different in colors and shades. Everything suggests something else to me, and from each thing comes yet another new thing.

Approaching toy design is, above all, learning to listen to and look at the beauty around us with the innocence, freshness, curiosity, and imagination of children. A daily exercise that stimulates creativity and from which a designer cannot escape.

Bibliography

Bang M. (1991), *Picture This,* Chronicle Books, San Francisco, 2016.
Huizinga J. (1938), *Homo ludens*, Einaudi, Turin, 2002.
Itten J. (1961), *Art of Color*, il Saggiatore, Milan, 2016.
Kandinsky W. (1968), *Point Line and Surface*, Adelphi, Milan, 2017.
Klee P. (1923), *Theory of Form and Figuration 1*, Mimesis, Milan-Udine, 2011.
Klee P. (1923), *Theory of Form and Figuration 2*, Mimesis, Milan-Udine, 2011.
Klee P. (1924), *Pedagogical Sketchbook*, Abscondita, Milan, 2002.
Manu A. (1995), *Tool Toys*, Danish Design Centre, Copenhagen, 1995.
Munari B. (1966), *Art as a Craft*, Editori Laterza, Rome-Bari, 1991.
Munari B. (1968), *Design and Visual Communication*, Editori Laterza, Rome-Bari, 1994.
Munari B. (1977), *Fantasy*, Editori Laterza, Rome-Bari, 1993.
Munari B. (1981), *From Thing to Thing*, Editori Laterza, Rome-Bari, 1992.
Munari B. (1960), *The Square*, Corraini, Mantua, 2017.
Munari B. (1964), *The Circle*, Corraini, Mantua, 2016.
Munari B. (1976), *The Triangle*, Corraini, Mantua, 2007.
Munari B. (1958), *Munari's Forks*, Corraini, Mantua, 1996.
Munari B. (1963), *Good Design*, Corraini, Mantua, 1998.
Murugarren M. (2003), *The Universal Bestiary of Professor Revillod*, Logos, Modena, 2010.
Pastoureau M. and Simonnet D. (2006), *The Little Book of Colors*, Ponte alle Grazie, Milan.
Restelli B. (2002), *Playing with Touch*, FrancoAngeli, Milan, 2007.
Rodari G. (1973), *Grammar of Fantasy*, Edizioni EL, Trieste, 1997.

I. Guerrieri, *Toy Design*, https://doi.org/10.1007/978-3-031-98872-1

Rodari G. (1979), "A mature gentleman with a green ear," in *Words to Play*, Manzuoli, Florence.

Ludography

Following the pattern of a bibliography, I wish to list here a few games that are significant to me for their history, for the formal research that distinguishes them, for the educational content they convey, and for the high-quality illustrations. The list does not follow the usual alphabetical order but instead a personal criterion that puts at the top of the list the games of two great authors, Enzo Mari and Bruno Munari, and then others grouped by type: ride-on games, construction games, memory games, classification games, balance games, magnetic games, interlocking games, and representation games. Not all have the author specified, because not all are identifiable; therefore, I found it more useful to indicate for each the website of the company to which they are associated.

The 16 Animals, Enzo Mari, danesemilano.com
The Little Monkey Zizi, Bruno Munari, lapermanentemobilicantu.com
Monkey, Kay Bojesen, kaybojesen-denmark.com
Crocodile, A.M. design office, plust.it
Rocker, Doshi Levien, richard-lampert.de
Checubo, Giovanni Fumagalli, gonzagarredi.com
My zoo, Martì Guixé, magisdesign.com
Polygrif, Moebe Design studio, moebe.dk
Quadrigo, Jean-Phillipe Rossinelli, naefspiele.ch
Kapla, kapla.com
FaceMaker, Miller Goodman, millergoodman.com
Memory Game, *Charley Harper*, ammobooks.com
Nomenclature, Valentina Gottardi, cocaidesign.it
Balancing Cactus, plantoys.com
The Gecko, beleduc.de

I. Guerrieri, *Toy Design*, https://doi.org/10.1007/978-3-031-98872-1

Magnetic Blocks, pixio.site
Mula Crane, Francis Cayouette, ikea.com
Magnetic Racer, tegu.com
Box with Fruit, Amalie Skov Rahbæk, flexaworld.com
Traffic Coasters, Irene Guerrieri, fratelliguzzini.com
Magic House, Giulia Serafin, headu.com
Thousand Theaters, Giulia Orecchia, erickson.it
Travel, Learn, Explore The Earth, Matteo Gaule, sassijunior.com
Dixit, Jean-Louis Roubira, libellud.com
First Orchard, Anneliese Farkaschovsky, haba.com
Green Bruco, Giorgio Vanetti and Loredana Farina, coccinella.com
From Flower to Fruit, Valentina Gottardi, cocaidesign.it

MIX
Papier aus verantwortungsvollen Quellen
Paper from responsible sources
FSC® C105338

If you have any concerns about our products,
you can contact us on
ProductSafety@springernature.com

In case Publisher is established outside the EU,
the EU authorized representative is:
Springer Nature Customer Service Center GmbH
Europaplatz 3, 69115 Heidelberg, Germany

Printed by Libri Plureos GmbH
in Hamburg, Germany